Polyoxometalate Chemistry

RSC Foundations

For a list of titles in this series see:
rsc.li/foundations

How to obtain future titles on publication:
A standing order plan is available for this series. A standing order will bring delivery of each new volume immediately on publication.

For further information please contact:
Book Sales Department, Royal Society of Chemistry, Thomas Graham House, Science Park, Milton Road, Cambridge, CB4 0WF, UK
Telephone: +44 (0)1223 420066, Fax: +44 (0)1223 420247
Email: booksales@rsc.org
Visit our website at books.rsc.org

Polyoxometalate Chemistry

By

Jan-Christian Raabe

University of Hamburg, Germany
Email: Jan-Christian.Raabe@Jan-Christian-Raabe.de

RSC Foundations No. 3

Paperback ISBN: 978-1-83707-203-3
EPUB ISBN: 978-1-83767-963-8
PDF ISBN: 978-1-83707-209-5
Print ISSN: 2978-1477
Electronic ISSN: 2977-0084

A catalogue record for this book is available from the British Library

The Royal Society of Chemistry is a charity, registered in England and Wales, Number 207890, and a company incorporated in England by Royal Charter (Registered No. RC000524), registered office: Burlington House, Piccadilly, London W1J 0BA, UK, Telephone: +44 (0)20 7437 8656.

For further information see our website at www.rsc.org

For general enquiries, please contact books@rsc.org

For EU product safety enquiries, please email books@rsc.org or contact Royal Society of Chemistry Worldwide (Germany) GmbH, Römischer Hof, Unter den Linden 10, 10117 Berlin.

Preface

Polyoxometalates – Small Versatile Nanoclusters

The discovery of polyoxometalates began in the 18th and 19th centuries. At this time, the experiments were simple and the molecular structures of these clusters were not clear. However, structural elucidations began in the early 20th century, using modern X-ray diffraction experiments. The first concrete applications were reported in the middle of the 20th century and were optimized in the 21st century. It took until the 21st century to recognize the potential of using these materials for current technologies. Since the 2010s, these materials have been used in sustainable and green chemistry processes and in the fight against different diseases (bacterial, viral, tumors, diabetes or Alzheimer's disease). This is possible because POMs are soluble in water, a green and environmentally friendly solvent. In addition, the solubility is easy to tailor for desired applications by choosing the correct cation. Due to the variability of available structure types and various elements that can be incorporated into the structure, POMs benefit from being customized for specific applications. As POMs can also be produced on a large scale, the first applications have already been implemented in the chemical industry, attracting more and more academic and industrial interest in POMs. The full potential of polyoxometalates has not yet been exhausted. These materials are therefore proving to be interesting candidates for opening further technologies and areas of application in order to prepare humanity for the challenges of this century.

Jan-Christian Raabe

RSC Foundations No. 3
Polyoxometalate Chemistry
By Jan-Christian Raabe

Published by the Royal Society of Chemistry, www.rsc.org

Acknowledgement

The author would like to gratefully thank:

- The publisher, the Royal Society of Chemistry, in particular Nicki Dennis, for the opportunity to write this book.
- His long-time friend Tim Grünther for his constant support and the shared moments.
- And of course, his friend Alexander Henning for so many years of friendship.

RSC Foundations No. 3
Polyoxometalate Chemistry
By Jan-Christian Raabe

Published by the Royal Society of Chemistry, www.rsc.org

Abbreviations

All abbreviations used in this book are listed here.

AA	Acetic acid
AChE	Acetylcholinesterase
AD	Alzheimer's disease
AIDS	Acquired immune deficiency syndrome
BChE	Butyrylcholinesterase
BO	Bond order
CMC	Carboxymethyl chitosan
COPD	Chronic obstructive pulmonary disease
CT	Charge transfer
CuAAC	Copper-catalyzed azide-alkyne cycloaddition
DCC	*N,N'*-Dicyclohexylcarbodiimide
DCM	Dichloromethane
DES	Deep eutectic solvents
DFT	Density functional theory
DMAP	Dimethyl amino pyridine
DMF	*N,N*-dimethylformamide
DNA	Deoxyribonucleic acid
EC_{50}	Half maximal effective concentration
ECODS	Extraction-coupled oxidative desulfurization
EDA	Ethyl diazoacetate
EPR	Electron spin resonance
FA	Formic acid
FluV A	Influenza virus
HBA	Hydrogen bond acceptor
HBD	Hydrogen bond donor

RSC Foundations No. 3
Polyoxometalate Chemistry
By Jan-Christian Raabe

Published by the Royal Society of Chemistry, www.rsc.org

HCV	Hepatitis C virus
HIV	Human immunodeficiency virus
HMF	Hydroxymethylfurfural
HPA	Heteropolyanion, heteropolyacid
HSAB	Hard and soft acids and bases
H_2en	Double protonated ethylendiamine
H_2tmen	Double protonated *N,N,N′,N′*-tetramethylethylene-diammonium
IC_{50}	Half-maximal inhibitory concentration
IL	Ionic liquid
imi	Imidazol
IPA	Isopolyanion, isopolyacid
IR	Infrared
IVCT	Intervalence charge transfer
LA	Levulinic acid
LLT	Ligand–ligand transfer
LMCT	Ligand-to-metal charge transfer
MLCT	Metal-to-ligand charge transfer
MO	Molecular orbital
MRSA	Methicillin-resistant *Staphylococcus aureus*
NMR	Nuclear magnetic resonance
NTPDases	Ecto-nucleoside triphosphate diphosphohydrolases
OA	Oxalic acid
ODS	Oxidative desulfurization
OxFA	Oxidative formation of formic acid
pH	Negative decadic logarithm of the hydrogen ion concentration
POM	Polyoxometalate
ppm	Parts per million
*p*TSA	*para*-Toluene sulfonic acid
RedOx	Reduction/oxidation
RNA	Ribonucleic acid
SAA	Sulfoacetic acid
2-SBA	2-Sulfobenzoic acid
sc-XRD	Single-crystal X-ray diffraction
2-SOBA	2-(Sulfooxy)benzoic acid
ss	Single-stranded
TBA	Tetrabutylammonium
TGA	Thermogravimetric analysis
THF	Tetrahydrofurane

TMSCN	Trimethylsilyl cyanide
TMSPOM	Transition metal-substituted POMs
TPPO	Triphenylphosphine oxide
UV-Vis	Ultraviolet-visible
VRSA	Vancomycin-resistant *Staphylococcus aureus*
XRD	X-ray diffraction

About the Author

Dr Jan-Christian Raabe studied chemistry at the University of Hamburg. During his PhD he worked in the field of polyoxometalate chemistry and developed new synthetic strategies for those cluster compounds. The polyoxometalates he developed were specifically designed for catalytic applications that can be done in water and that convert biomass substrates from which industrially relevant platform chemicals can be produced. His work makes a significant contribution to the field of green chemistry. Some of his new polyoxometalates were successfully tested in catalytic applications. The results of his research have been published in internationally renowned journals.

RSC Foundations No. 3
Polyoxometalate Chemistry
By Jan-Christian Raabe

Published by the Royal Society of Chemistry, www.rsc.org

About This Book

The book addresses the following topics:

- A short introduction into the topic of polyoxometalate chemistry (structural chemistry).
- Basic principles of coordination chemistry, especially oxo chemistry.
- Synthetic inorganic chemistry.
- Principles of inorganic chemistry related to the chemistry of polyoxometalates.
- Principles of catalysis, reaction technology and different catalytic applications for polyoxometalates.
- Applications of polyoxometalates in biomedicine (for the treatment of viruses, multidrug-resistant bacteria, tumors and diabetes) and fundamental principles of biochemistry in order to understand the interaction between polyoxometalates and biomolecules.

The topic of polyoxometalates is rarely addressed in inorganic chemistry textbooks. However, the chemistry of polyoxometalates shows many parallels between the concepts of coordination chemistry and the fundamental concepts of inorganic chemistry. Therefore, it is not a simple niche topic, but a fully established complex of topics in inorganic chemistry.

The biochemical basics for understanding the chapter "Applications of Polyoxometalates in Biomedicine" (Chapter 6) are only briefly covered. For a more detailed understanding, readers should consult biochemistry textbooks, for example:

1. S. Habtemariam, *Basic Chemistry for Life Science Students and Professionals*, The Royal Society of Chemistry, 2023.

RSC Foundations No. 3
Polyoxometalate Chemistry
By Jan-Christian Raabe

Published by the Royal Society of Chemistry, www.rsc.org

2. *Molecular Biology and Biotechnology*, ed. R. Rapley, The Royal Society of Chemistry, 2021.
3. M. Sinnott, *Carbohydrate Chemistry and Biochemistry: Structure and Mechanism*, The Royal Society of Chemistry, 2013.

It should be noted that this book cannot present every single application and every single POM molecule. For further applications, the reader is referred to the literature.

This book is supported by 75 figures, 8 tables, 3 schemes and many equations.

Who Should Read This Book?

This book should be read by anyone working in the field of polyoxometalate chemistry. The book addresses readers who are involved in research in other fields of inorganic chemistry in order to expand their own horizons and gain new impressions for their own work. It is also intended for anyone with an interest in inorganic chemistry or polyoxometalate chemistry, such as students or private citizens.

It should be noted that the level of this book is advanced and requires some basic knowledge of chemistry. However, the topic is also addressed for readers who are interested in fundamental concepts of inorganic chemistry and want to understand these concepts better.

RSC Foundations No. 3
Polyoxometalate Chemistry
By Jan-Christian Raabe

Published by the Royal Society of Chemistry, www.rsc.org

Note Added After First Publication

Since the original publication of this book online and as an EPUB in October 2025, amendments have been made to correct errors in the original and improve the order of content within chapters. These changes are reflected in all versions available from the print date of March 2026.

RSC Foundations No. 3
Polyoxometalate Chemistry
By Jan-Christian Raabe

Published by the Royal Society of Chemistry, www.rsc.org

Contents

RSC Foundations No. 3
Polyoxometalate Chemistry
By Jan-Christian Raabe

Published by the Royal Society of Chemistry, www.rsc.org

1 Introduction

Polyoxometalates (POMs) are so-called inorganic cluster compounds with highly anionic charges. Formally, POMs are metal oxo complexes (Figure 1.1) formed with elements of group five (vanadium V, niobium Nb and tantalum Ta) and six (molybdenum Mo and tungsten W) of the periodic table, mostly in their highest oxidation states of plus five (group five) and plus six (group six). Oxo ligands can be coordinated terminally (M=O) or bridging (M–O–M). The metals are coordinated by six oxo ligands in an octahedral coordination motif (Figure 1.1), so the POMs are formed by the connection of different metal–oxygen octahedra (MO_6) by sharing edges or corners.[1–7]

Connecting the octahedra by surfaces is not preferred, due to the high cationic repulsion that results from the high cationic charges from the metal atoms.[7]

In the field of inorganic chemistry, the topic of POM chemistry can be classified between molecular and solid-state chemistry. Molecules that belong to the class of molecular chemistry are small inorganic molecules with a defined symmetry, a defined number of atoms and are mostly highly soluble in aqueous media. On the other side, metal oxides belong to the class of solid-state chemistry. Formally, oxides are infinitely expanded clusters with almost an infinite number of M–O–M bridges. From this point of view, POMs can be classified as examples from solid-state chemistry. However, a classification of molecular chemistry is also correct, because POMs are also defined clusters with a defined number of atoms, with a defined symmetry and are mostly highly soluble in aqueous media. The mentioned points are summarized in Figure 1.2.[8]

RSC Foundations No. 3
Polyoxometalate Chemistry
By Jan-Christian Raabe

Published by the Royal Society of Chemistry, www.rsc.org

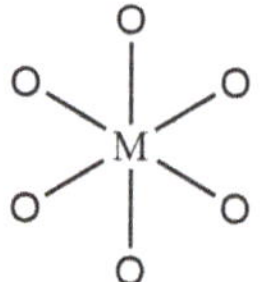

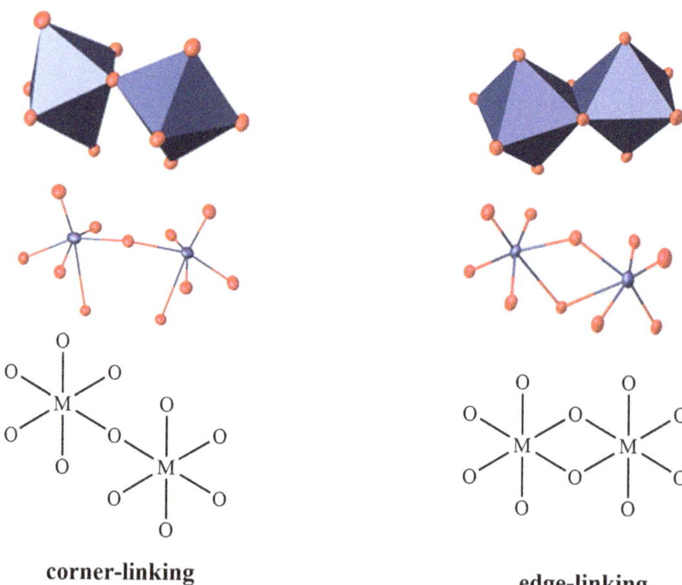

Figure 1.1 A metal oxo complex (top). A metal M is coordinated by six oxo ligands, forming a MO_6 octahedron. Two octahedra can connect with each other by common corners (left) or edges (right).

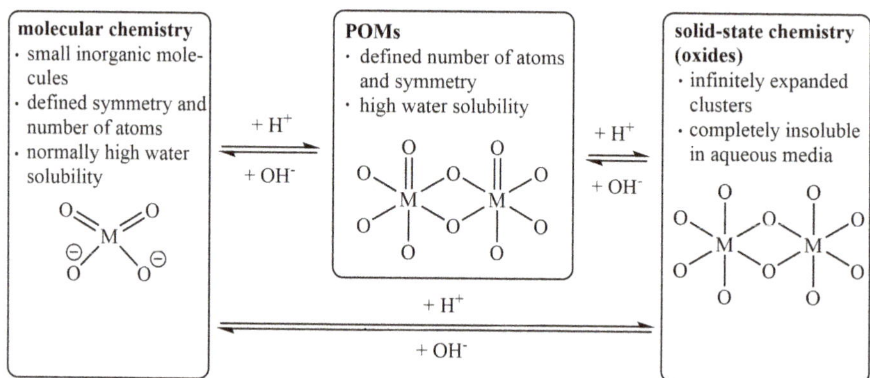

Figure 1.2 POM chemistry as a special topic between molecular and solid-state chemistry. Small inorganic oxo compounds can be converted in acid media to POMs or metal oxides. POMs and metal oxides can be converted to oxo compounds in basic pH media. It is also possible to convert metal oxides to POMs in basic media.

Examples of the different reaction types are shown in Figure 1.2 and eqn (1.1)–(1.6).

- The conversion of sodium tungstate Na_2WO_4 or sodium molybdate Na_2MoO_4 to tungsten oxide WO_3 and molybdenum oxide MoO_3 in acidic pH media.[1] Eqn (1.1) shows the acidification of MO_4^{2-} to oxides:

$$\begin{aligned} Na_2WO_4 + 2HCl &\rightarrow WO_3 + H_2O + 2NaCl \\ Na_2MoO_4 + 2HCl &\rightarrow MoO_3 + H_2O + 2NaCl \end{aligned} \tag{1.1}$$

- The conversion of small inorganic compounds to POMs in acidic pH media.[1,9,10] Eqn (1.2) shows the acidification of MO_4^{2-} to POMs:

$$\begin{aligned} 6Na_2WO_4 + 10HCl &\rightarrow Na_2[W_6O_{19}] + 5H_2O + 10NaCl \\ 6Na_2MoO_4 + 10HCl &\rightarrow Na_2[Mo_6O_{19}] + 5H_2O + 10NaCl \end{aligned} \tag{1.2}$$

- The conversion of POMs into oxides in acidic media.[1] Eqn (1.3) shows the acidification of POMs to oxides:

$$\begin{aligned} Na_2[W_6O_{19}] + 2HCl &\rightarrow 6WO_3 + H_2O + 2NaCl \\ Na_2[Mo_6O_{19}] + 2HCl &\rightarrow 6MoO_3 + H_2O + 2NaCl \end{aligned} \tag{1.3}$$

- The conversion of oxides to POMs in basic pH media.[1,11,12] Eqn (1.4) shows a general example for the basic degradation of oxides to POMs:

$$\begin{aligned} 3Nb_2O_5 + 8KOH &\rightarrow K_8[Nb_6O_{19}] + 4H_2O \\ 3Ta_2O_5 + 8KOH &\rightarrow K_8[Ta_6O_{19}] + 4H_2O \end{aligned} \tag{1.4}$$

- The conversion of POMs in basic pH media to small inorganic oxo compounds.[1] Eqn (1.5) shows a general reaction equation for the basic degradation of POMs to MO_4^{2-}:

$$\begin{aligned} Na_2[W_6O_{19}] + 10NaOH &\rightarrow 6Na_2WO_4 + 5H_2O \\ Na_2[Mo_6O_{19}] + 10NaOH &\rightarrow 6Na_2MoO_4 + H_2O \end{aligned} \tag{1.5}$$

- The conversion of metal oxides in basic pH media to oxo compounds.[1] Eqn (1.6) shows a general example for the basic degradation of oxides to MO_4^{2-}:

$$\begin{aligned} WO_3 + 2NaOH &\rightarrow Na_2WO_4 + H_2O \\ MoO_3 + 2NaOH &\rightarrow Na_2MoO_4 + H_2O \end{aligned} \tag{1.6}$$

This is the reason why so many oxides are sensitive against hydroxides. A very well known example is the basic degradation of glass (SiO_2). This becomes noticeable because the glass turns cloudy when it is exposed to hydroxide for a long time.[1,8]

The different conversion reactions highlight the synthetic potential of POM chemistry and the fundamental position of POM chemistry in the field of inorganic chemistry. The fundamental reaction principle involves a monomeric species MO_4^{2-} dimerizing (or oligomerizing) under acidic conditions by splitting off one (or more) molecule(s) of water. Under basic conditions the dimer (or the oligomer) is split down to its monomers MO_4^{2-}. Eqn (1.7) is an example for a dimerization/oligomerization of MO_4^{2-} under water elimination:[1,8]

$$2MO_4^{2-} \underset{+OH^-}{\overset{+H^+}{\rightleftarrows}} [M_2O_7]^{2-} + H_2O \tag{1.7}$$

In general, there are two known classes of POMs:

- POMs that contain an additional main-group heteroelement polyhedron XO_y. This class is known as the heteropolyanion or heteropolyacid (HPA) class. Here, the heteroelement is mostly a main-group element of the periodic table like silicon Si, phosphorus P, antimony Sb or tellurium Te.
- POMs without an additional heteroelement are classified as isopolyanion or isopolyacid (IPA) structures.

Examples for these two classes will be discussed in the next part.

There are different possible ways of connecting the different MO_6 octahedra and the XO_y polyhedra. This is the reason there are so many known structure types for POMs. In 1826, Jöns Jakob Berzelius was able to synthesize the so-called phosphomolybdate anion with the general molecular composition of $[PMo_{12}O_{40}]^{3-}$.[13] Today, this structure is known as the Keggin-type structure with the general molecular stoichiometry of $[XM_{12}O_{40}]^{n-}$.

In 1864, Jean-Charles Galissard de Marignac was able to synthesize silicotungstate $[SiW_{12}O_{40}]^{4-}$ for the first time.[14] But there were no ideas about the structural properties of these compounds. With the work of Arthur Rosenheim and Otto Liebknecht in the beginning of the 20th century, the first hypotheses about the structural properties were born. It was assumed that the clusters were polyanionic structures with octahedra as fundamental building blocks, which were connected by corners and edges.[15–18] In 1934, James Fargher Keggin was able to elucidate the structure of the phosphotungstate anion $[PW_{12}O_{40}]^{3-}$ (the W analogue of $[PMo_{12}O_{40}]^{3-}$), which was named the Keggin-type structure.[19] The three-dimensional structure of a Keggin-type cluster is shown in Figure 1.3 in atomistic and polyhedron representations. A Keggin-type structure belongs to the class of HPA structures.[7]

With the use of modern X-ray diffraction experiments, scientists were able to elucidate the complete three-dimensional structure of different POMs. So, the Keggin-type structure was elucidated, where twelve MO_6 octahedra are connected around one XO_4 tetrahedron, in which each of the four oxygen atoms is part of a M_3O_{13} triad.

Another very popular structure is the HPA Wells–Dawson-type structure, with the general molecular formula $[X_2M_{18}O_{62}]^{n-}$. The first synthesized Wells–Dawson-type cluster was the compound $K_6[P_2W_{18}O_{62}]$, synthesized by Friedrich Kehrmann in 1894.[20] At this time, the compound had the empirical formula $3K_2O{\cdot}P_2O_5{\cdot}18WO_3{\cdot}14H_2O$, which was verified by Rosenheim and Johannes Jaenicke in 1917.[20–23] The real three-dimensional structure was suggested by Alexander Frank Wells[23] and later verified by Barrie Dawson[22] in 1953 with single-crystal X-ray diffraction (sc-XRD), using 18-tungstophosphate $[P_2W_{18}O_{62}]^{6-}$ as an example. Here, the compound was isolated as salt $K_6[P_2W_{18}O_{62}]{\cdot}14H_2O$.[24,25] In the Wells–Dawson-type structure there are two XO_4 tetrahedrons, each coordinating a M_3O_{13} triad and one M_6O_{14} belt.[7] The three-dimensional structure type is shown in Figure 1.4 in both types of representation.

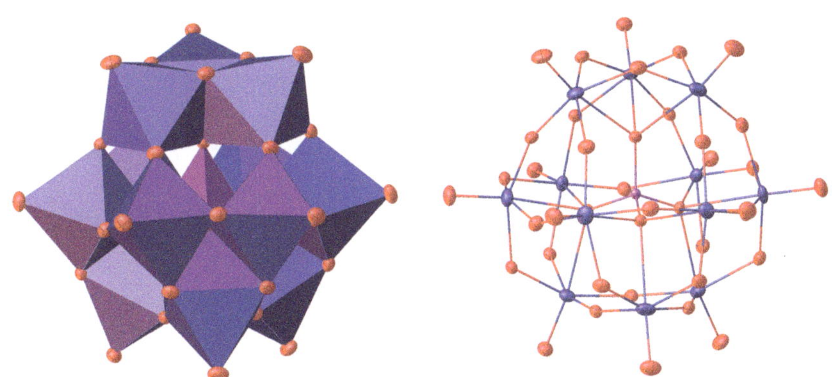

Figure 1.3 The Keggin-type structure $[XM_{12}O_{40}]^{n-}$. Polyhedron (left) and atomistic representation (right). Color code: purple – X, blue – metals M, and red – O. The data were used from the Cambridge Crystallographic Data Centre and Fachinformationszentrum Karlsruhe Access Structures service database (deposition number: 2177881).

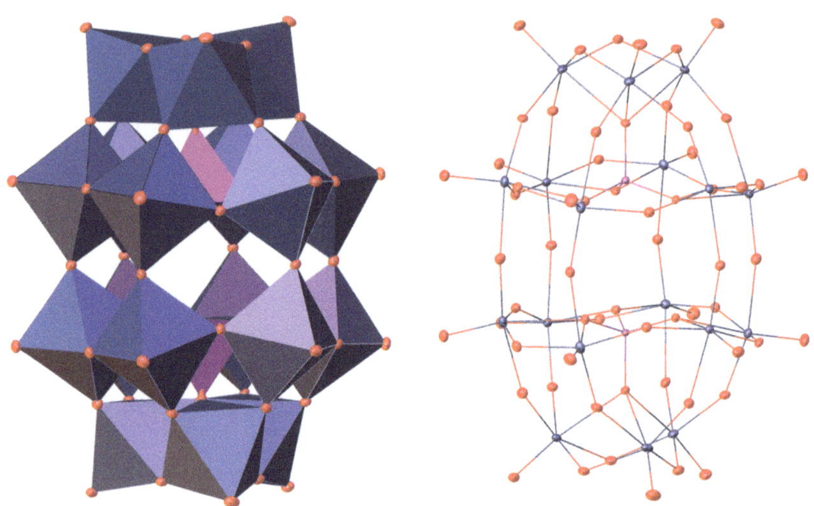

Figure 1.4 The Wells–Dawson-type structure $[X_2M_{18}O_{62}]^{n-}$. Polyhedron (left) and atomistic representation (right). Color code: purple – X, blue – metals M, and red – O. The data were used from the Cambridge Crystallographic Data Centre and Fachinformationszentrum Karlsruhe Access Structures service database (deposition number: 2216946).

Another type of structure, belonging to the class of HPA structures, is the Anderson–Evans-type with the general molecular formula $[XM_6O_{24}]^{n-}$.[7] This structure type is planar. The heteroelement X is in octahedral coordination with six oxygen atoms and can be represented by Te or Sb.[26–29] This planar geometry was first postulated by John Stuart Anderson in 1937[30] and later verified by Howard Tasker Evans in 1948 using sc-XRD for the anion $[TeMo_6O_{24}]^{6-}$, which was isolated as ammonium NH_4^+ and potassium K^+ salt.[31] The three-dimensional structure type is shown in Figure 1.5 in both representations.

An example of an IPA structure type is the Lindqvist-type structure with the general molecular formula $[M_6O_{19}]^{n-}$.[11] This anion is also called hexametalate. The namesake for this structure was the Swedish chemist Fritz Ingvar Lindqvist, who focused his research on isopolymolybdates in 1950.[32] In 1953 he was able to verify this structure type for the anion $[Nb_6O_{19}]^{8-}$, called hexaniobate. One year later (1954) the proof for the anion $[Ta_6O_{19}]^{8-}$ (hexatantalate) was provided in cooperation with Aronsson.[33] POMs of the Lindqvist-type are known for elements of group 5 and 6 of the periodic table in their highest oxidation states.[34] All six MO_6 octahedrons are connected by sharing corners and edges to form a larger octahedron. Here, the octahedra are arranged around a central, sixfold and octahedrally

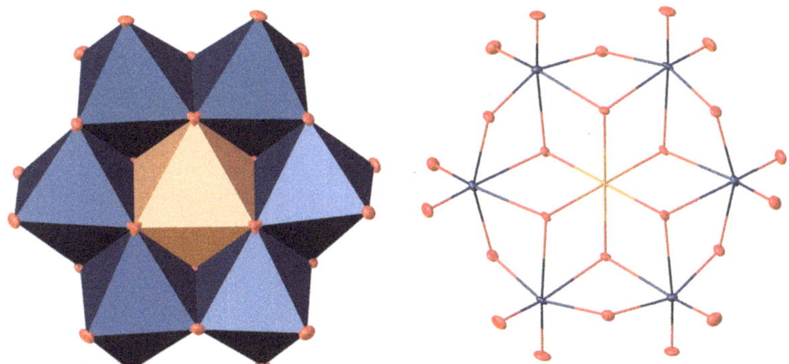

Figure 1.5 The Anderson–Evans-type structure $[XM_6O_{24}]^{n-}$. Polyhedron (left) and atomistic representation (right). Color code: orange – X, blue – metals M, and red – O. The data were used from the Cambridge Crystallographic Data Centre and Fachinformationszentrum Karlsruhe Access Structures service database (deposition number: 2293848).

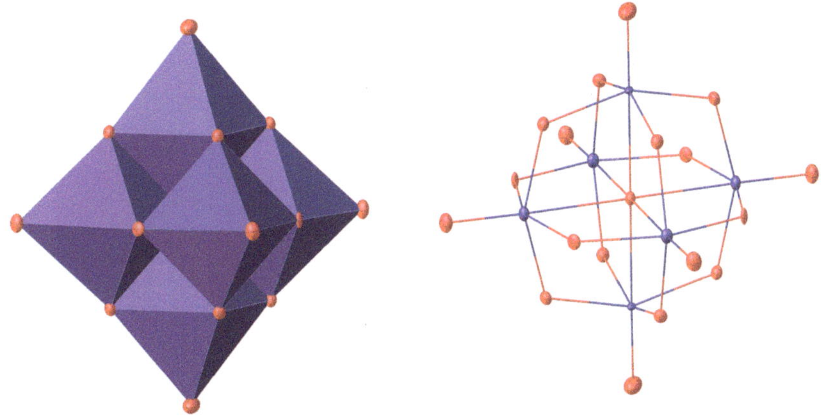

Figure 1.6 The Lindqvist-type structure $[M_6O_{19}]^{n-}$. Polyhedron (left) and atomistic representation (right). Color code: blue – metals M and red – O. The data were used from the Cambridge Crystallographic Data Centre and Fachinformationszentrum Karlsruhe Access Structures service database (deposition number: 2220347).

coordinated oxo ligand μ_6-O, with each of the six metals carrying a terminally coordinated oxo ligand $M{=}O_t$, which fills the corners of the hexametalate octahedron.[33] The three-dimensional structure of the hexametalate Lindqvist-type structure is shown in both representations in Figure 1.6.

Lindqvist-type POMs have already been interpreted as so-called clathrate (inclusion) compounds based on the host/guest concept. The

μ_6-O oxo ligand has the role of the guest, which gets the full anionic charge, while the metal oxide shell, the host, is considered to be electroneutral. This hypothesis is strengthened by crystallographic data, suggesting a significantly extended bond length for the μ_6-O metal bonds compared to the bond lengths for the metal–metal bridging oxo ligands μ_2-O.[33,35,36] Using density functional theory (DFT) based calculations, it was shown that the cluster $\{M_6O_{18}\}$ shows an enhanced stability without the μ_6-O ligand, strengthening the role of the host for the $\{M_6O_{18}\}$ cluster in relation to the host/guest hypothesis.[33,37]

Of all POM structure types, lacunary-type structures have a special role. Lacunary-type structures are not an independent structure type. These structures are derived from any structure type (Keggin-, Wells–Dawson-type) in which one or more MO_x octahedra have been removed.[38–41] The resulting defects or vacancies make lacunary-type structures particularly reactive towards foreign metal ions or towards themselves. For example, a dimerization of the Keggin lacunary-type anion $[PMo_9O_{34}]^{9-}$ to the Wells–Dawson-type structure $[P_2Mo_{18}O_{62}]^{6-}$ is known.[42] However, the reactivity of lacunary-type structures can also be used for synthetic approaches, as will be shown in later chapters. A three-dimensional representation of a Keggin lacunary-type structure is shown in Figure 1.7. Lacunary oxygen atoms are oxo ligands that form the vacancies which are normally connected to the metals in their intact structure type.

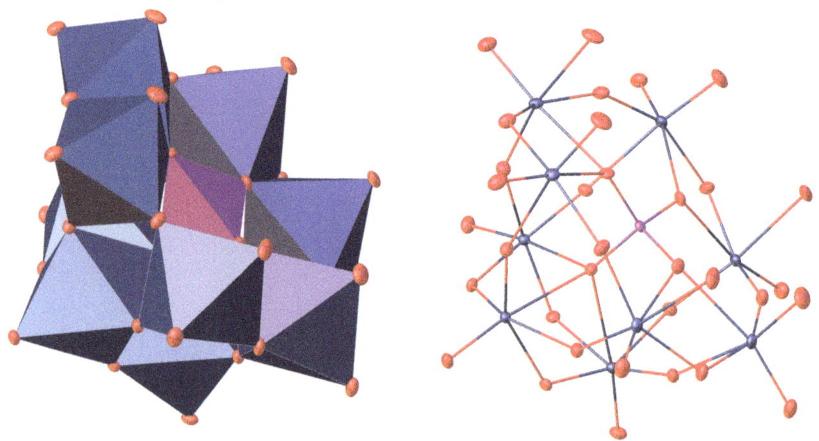

Figure 1.7 The Keggin lacunary-type structure $[XM_9O_{34}]^{n-}$. Polyhedron (left) and atomistic representation (right). Color code: purple – X, blue – metals M, and red – O. The data were used from the Cambridge Crystallographic Data Centre and Fachinformations-zentrum Karlsruhe Access Structures service database (deposition number: 2205006).

In summary, the structure types discussed above are the most important ones. There are also some less known structure types:

- Dexter–Silverton-type structure $[XM_{12}O_{42}]^{n-}$. This type of structure was investigated in 1968 by David D. Dexter and James V. Silverton and was synthesized for the first time by Louis C. W. Baker, George A. Gallagher and Thomas P. McCutcheon in 1953.[43] It was first reported for the anion $[CeMo_{12}O_{42}]^{8-}$, which was isolated as $(NH_4)_2H_6[CeMo_{12}O_{42}]\cdot 12H_2O$ salt.[7,44] Here, the element cerium (oxidation state +4) acts as a heteroelement, which is coordinated in an icosahedron CeO_{12} coordination motif. The MoO_6 octahedra are connected *via* face sharing, which is not often found in POM chemistry, as the highly charged metal cations repel each other. This type of structure belongs to the class of HPA.[7]
- Waugh-type structure $[XM_9O_{32}]^{n-}$. This structure type was first reported in 1954 by John L. T. Waugh using the anion $[MnMo_9O_{32}]^{6-}$ as an example, the so-called 9-molybdomanganate. It was first synthesized by Hall in 1907.[45,46] The compound was originally isolated as $(NH_4)_6[MnMo_9O_{32}]\cdot 8H_2O$ salt. This type of structure belongs to the class of HPA.[47] The heteroelement manganese Mn (oxidation state +4) is octahedrally coordinated by six oxygen atoms, as shown in Figure 1.8.[48] Each of the six oxo ligands of the central MnO_6 octahedron coordinates to three metal centers. The IPA analogue of the Waugh-type structure is the anion $[Mo_{10}O_{32}]^{4-}$ and is called decamolybdate.[49]

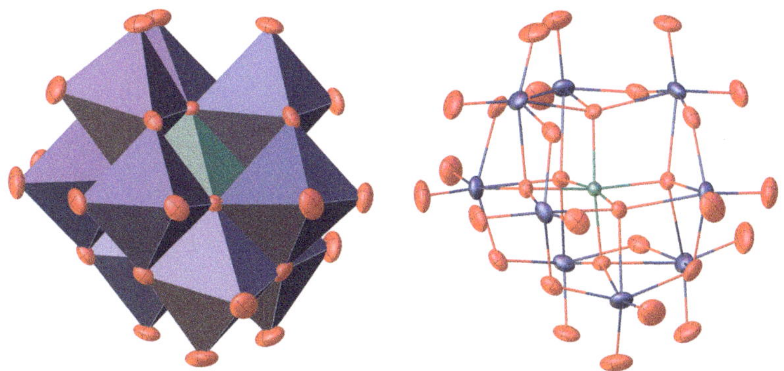

Figure 1.8 The Waugh-type structure $[XM_9O_{32}]^{n-}$. Polyhedron (left) and atomistic representation (right). Color code: green – X, blue – metals M, and red – O. The data were used from the Cambridge Crystallographic Data Centre and Fachinformationszentrum Karlsruhe Access Structures service database (deposition number: 728250).

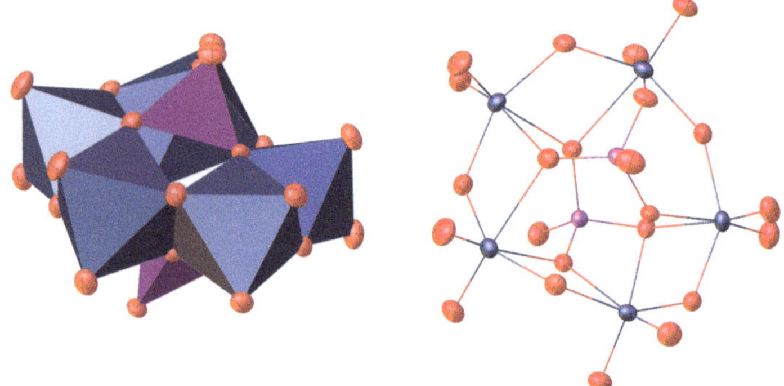

Figure 1.9 The Strandberg-type structure $[X_2M_5O_{23}]^{n-}$. Polyhedron (left) and atomistic representation (right). Color code: purple – X, blue – metals M, and red – O. The data were used from the Cambridge Crystallographic Data Centre and Fachinformationszentrum Karlsruhe Access Structures service database (deposition number: 803421).

- Strandberg-type structure $[X_2M_5O_{23}]^{n-}$. This type of structure was discovered in 1973 by Rolf Strandberg for the anion $[P_2Mo_5O_{23}]^{6-}$, which was isolated as $Na_6[P_2Mo_5O_{23}]\cdot13H_2O$ salt and belongs to the class of HPA structures.[50] The typical three-dimensional structure of the Strandberg-type anion is shown in Figure 1.9. There are two PO_4 tetrahedrons on the top and bottom of the metalate framework, which are connected by corners to the MO_6 octahedrons. Each of the framework metals is coordinated to both PO_4 tetrahedrons, resulting in a PO–M–OP bond motif. All five edge sharing MO_6 octahedrons are found in a pentagonal arrangement. One oxygen atom of each PO_4 tetrahedron is oriented outside the cluster with no coordination partner.
- Anion $[P_4W_{14}O_{58}]^{12-}$. The anion $[P_4W_{14}O_{58}]^{12-}$ is a widely unknown HPA motif, which has not yet been much reported in literature.[51,52] It was first synthesized by Kehrmann[53] in 1892, but at this time there was no idea about the stoichiometric element ratios and the structure. In a study from 2024, it was shown that this anion is formed by ${WO_4}^{2-}$ and ${PO_4}^{3-}$ at pH values of 5 in aqueous solutions. It is an intermediate in the formation of a Keggin-type structure $[PW_{12}O_{40}]^{3-}$ (at pH values of 1). Structurally, there are two oppositely oriented Keggin-type hemispheres that are connected by two corner sharing PO_4 tetrahedrons. Both hemispheres consist of seven WO_6 octahedrons, each with a central PO_4 tetrahedron.[54] The three-dimensional structure is shown in Figure 1.10.

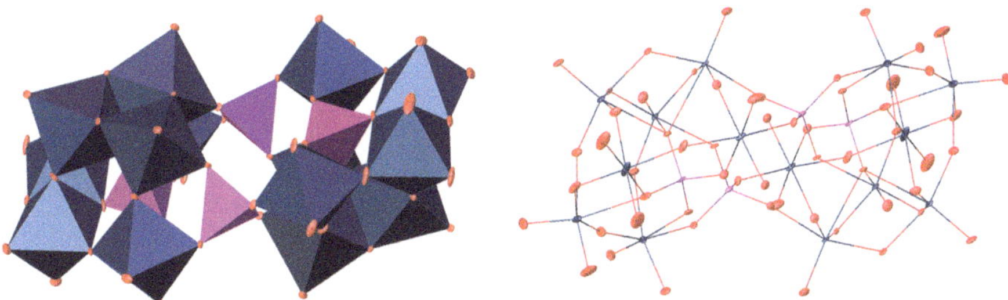

Figure 1.10 The $[P_4W_{14}O_{58}]^{12-}$ structure. Polyhedron (left) and atomistic representation (right). Color code: purple – P, blue – W, and red – O. The data were used from the Cambridge Crystallographic Data Centre and Fachinformationszentrum Karlsruhe Access Structures service database (deposition number: 2293849).

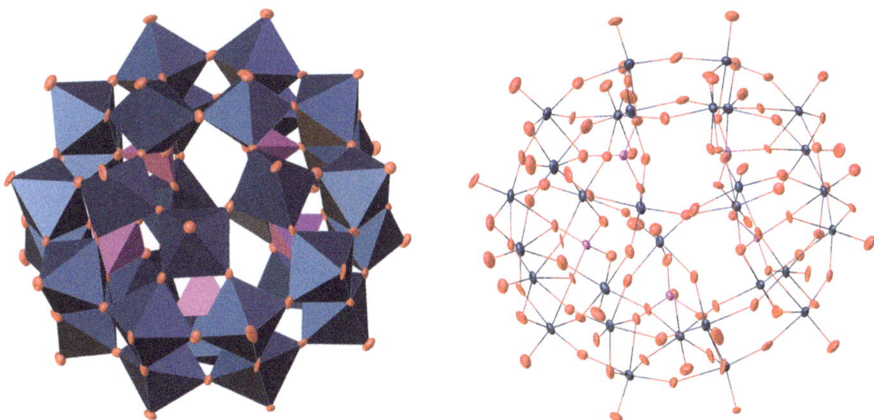

Figure 1.11 The Preyssler-type structure $[P_5W_{30}O_{110}]^{15-}$. Polyhedron (left) and atomistic representation (right). Color code: purple – P, blue – W, and red – O. The data were used from the Cambridge Crystallographic Data Centre and Fachinformationszentrum Karlsruhe Access Structures service database (deposition number: 1014312).

- Preyssler-type structure $[P_5W_{30}O_{110}]^{15-}$. This HPA structure was discovered in 1970 by C. Preyssler.[55] The anion $[P_5W_{30}O_{110}]^{15-}$ contains five PO_4 tetrahedrons, 30 WO_6 octahedrons and a large, pentagonal cavity in the center, as shown in Figure 1.11. All MO_6 and PO_4 polyhedra are connected *via* corner sharing in a ring-like structure. Formally, a Preyssler-type structure can be considered as a condensation product of five Keggin $\{PW_6O_{22}\}^{3-}$ units.[55,56] In this type of structure, the cavity can be used to complex different valent metal cations in order to use Preyssler-type POMs as caging ligands. This will be discussed in Chapter 4, Section 4.1.1.
- 14-Heteropolyvanadate $[PV_{14}O_{42}]^{9-}$. The stoichiometric composition of this HPA structure was determined by P. Souchay and

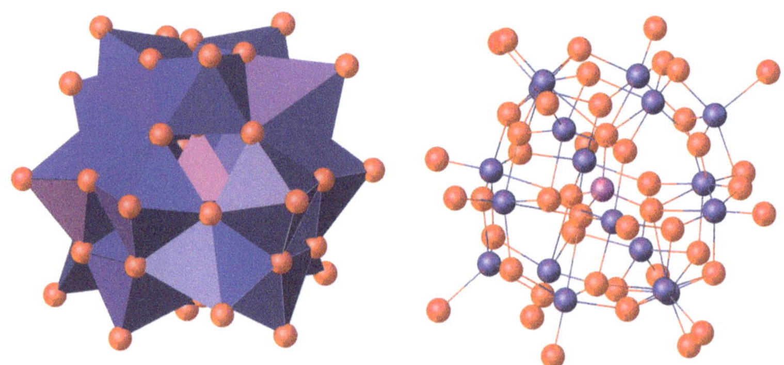

Figure 1.12 The 14-heteropolyvanadate $[PV_{14}O_{42}]^{9-}$ structure. Polyhedron (left) and atomistic representation (right). Color code: purple – P, blue – V, and red – O. The data were used from the Cambridge Crystallographic Data Centre and Fachinformationszentrum Karlsruhe Access Structures service database (deposition number: 1170424).

S. Dubois as $[PV_{12}O_{36}]^{7-}$. However, this result was doubted by the scientists F. Preuss and H. Schug. Instead, they proposed a P:V stoichiometry of 1:13 or 1:14. R. Kato, A. Kobayashi and Y. Sasaki successfully verified the element stoichiometry as 1:14. The structure is very similar to the Keggin-type structure. On a typical Keggin-type structure there are 14 so-called "pits", which provide further coordination sites for metal cations. The remaining two VO^{2+} groups (vanadyl cations) occupy *trans*-located sites and form a bipyramidal cap, as shown in Figure 1.12.[57] The concept of the pits on a Keggin-type structure will be discussed later in more detail.

- Decatungstate $[W_{10}O_{32}]^{4-}$ is an IPA motif and was reported by Oskar Max Glemser, W. Holznagel, W. Höltje and E. Schwarzmann in 1965.[58] First, it was assumed that the K^+ salt has the stoichiometric composition $K_5H[W_{12}O_{39}]$ and was therefore called polytungstate Y.[58,59] It was later shown by Ekkehard Birkholz, Joachim Fuchs, Wolfgang Schiller and Hans-Peter Stock that the real stoichiometric composition is $K_4[W_{10}O_{32}]$. So, the polytungstate Y is a decatungstate and can be interpreted as a dimer of a hypothetical Lindqvist lacunary-type structure $[M_5O_{16}]^{n-}$. Both $[M_5O_{16}]$ units are connected by four corner sharing MO_6 octahedra, generating a small void inside the cluster. The central oxo ligand is coordinated to only five metal atoms instead of six metal atoms in a Lindqvist-type structure. This structure-type is an intermediate during the acidification of aqueous WO_4^{2-} solutions at pH values around 2 to prepare WO_3.[59] The potential of decatungstate as a photocatalyst in photochemical applications has

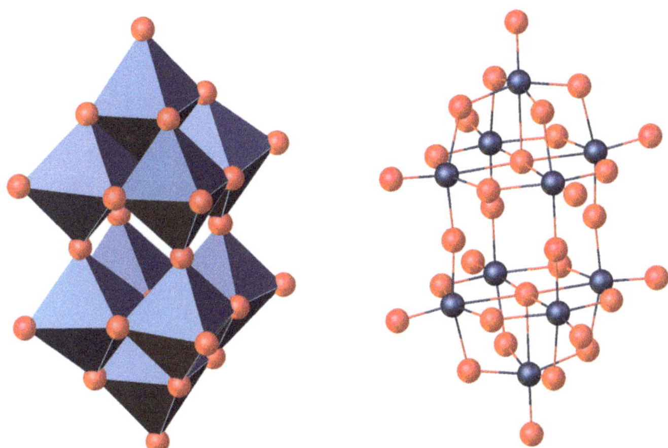

Figure 1.13 The decatungstate structure $[W_{10}O_{32}]^{4-}$. Polyhedron (left) and atomistic representation (right). Color code: blue – W and red – O. The data were used from the Cambridge Crystallographic Data Centre and Fachinformationszentrum Karlsruhe Access Structures service database (deposition number: 1115723).

been explored since the end of the 1970s.[60–62] A three-dimensional structure is shown in Figure 1.13.

- Weakley–Yamase-type structure $[L(W_5O_{18})_2]^{9-}$. This type of structure is an IPA motif, which can be understood as a dimer of two hypothetical Lindqvist lacunary-type structures $[W_5O_{18}]^{6-}$. The two lacunary-type structures use their vacant oxygen atoms to coordinate with a lanthanide cation L in its oxidation state of +3, which is surrounded by eight oxo ligands in a twisted cube-like arrangement. So, this structure type can be interpreted as a sandwich-type structure. The discovery goes back to the work of Timothy J. R. Weakley and Toshihiro Yamase, who reported compounds like $[Ce(W_5O_{18})_2]^{9-}$, $[Eu(W_5O_{18})_2]^{9-}$ and $[Tb(W_5O_{18})_2]^{9-}$.[63–69] A schematic representation is shown in Figure 1.14.
- Decavanadate $[V_{10}O_{28}]^{6-}$. This structure is an IPA motif. $[V_{10}O_{28}]^{6-}$ can be interpreted as a dimeric Lindqvist-type species, where two MO_6 octahedra are shared on one edge, as shown in Figure 1.15. It is an intermediate during the acidification of aqueous (*meta-*) vanadate VO_3^- solutions to prepare divanadium pentoxide V_2O_5. Comparable to a Lindqvist-type anion, all ten VO_6 octahedra are connected *via* edge sharing. From this anion different salts are known, *e.g.* a calcium (called Pascoite), 1,6-hexanediammonium or sodium Na^+ salt. It is also possible to isolate the POM acid $H_6[V_{10}O_{28}]$.[70–72] The POM acid has synthetic uses, which will be discussed in the following chapters.[73,74]

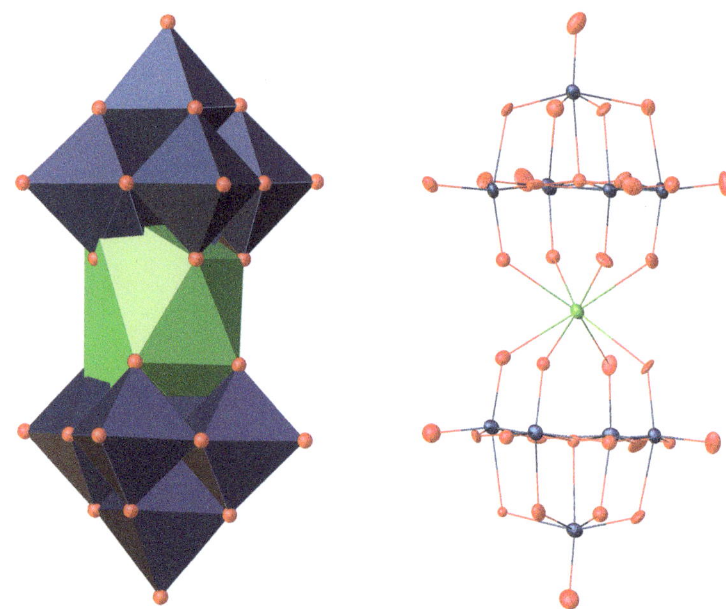

Figure 1.14 The Weakley-Yamase structure $[L(W_5O_{18})]^{9-}$. Polyhedron (left) and atomistic representation (right). Color code: green - L (lanthanides), blue - Mo and red - O. The data were used from the Cambridge Crystallographic Data Centre and Fachinformationszentrum Karlsruhe Access Structures service database (deposition number: 1400235).

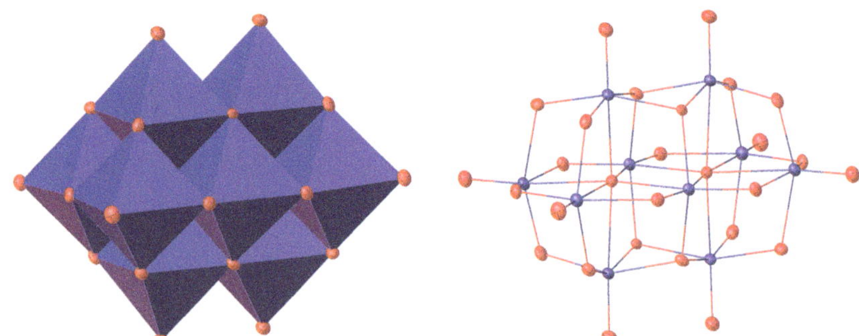

Figure 1.15 The decavanadate structure $[V_{10}O_{28}]^{6-}$. Polyhedron (left) and atomistic representation (right). Color code: blue - V and red - O. The data were used from the Cambridge Crystallographic Data Centre and Fachinformationszentrum Karlsruhe Access Structures service database (deposition number: 163138).

- Paradodecatungstate $[W_{12}O_{40}]^{8-}$ is an IPA motif and is also known as paratungstate B. The three-dimensional motif is shown in Figure 1.16. The anion has a big void in the center and can be

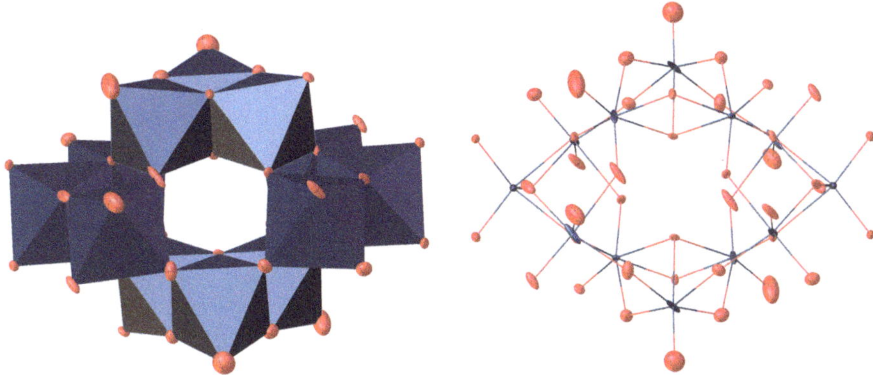

Figure 1.16 The paradodecatungstate or paratungstate B structure $[W_{12}O_{40}]^{8-}$. Polyhedron (left) and atomistic representation (right). Color code: blue – W and red – O. The data were used from the Cambridge Crystallographic Data Centre and Fachinformationszentrum Karlsruhe Access Structures service database (deposition number: 2293850).

understood as a Keggin-type structure without a heteroelement. This structure is an intermediate during acidification of aqueous WO_4^{2-} solutions to WO_3, which forms at pH values around 5.[54] It was first reported by Howard Tasker Evans and Orville W. Rollins in 1976.[75] Further investigations were done in the same year by Janina Chojnacka, Ewa Hodorowicz and Stanislaw Sagnowski.[76]

- Paratungstate $[W_7O_{24}]^{6-}$ is an IPA motif and is also known as paratungstate A. The existence of this simple anion in aqueous solution is only postulated. Attempts to isolate this anion from aqueous solution led to the transformation of the structure type to paradodecatungstate $[W_{12}O_{40}]^{8-}$. However, the anion $[W_7O_{24}]^{6-}$ can be isolated from aqueous solutions by precipitation with extremely complex cations. This means that there is a chemical equilibrium between both species $[W_7O_{24}]^{6-}$ and $[W_{12}O_{40}]^{8-}$, which shifts to the side of $[W_{12}O_{40}]^{8-}$ during crystallization. A three-dimensional representation of this anion is shown in Figure 1.17. All seven WO_6 octahedra are connected *via* edges in a butterfly arrangement.[77–80]
- Paramolybdate $[Mo_7O_{24}]^{6-}$ is an IPA motif, which has been known since the 1930s and is also called heptamolybdate. The most popular compound is ammonium heptamolybdate $(NH_4)_6[Mo_7O_{24}]$. A three-dimensional structure is shown in Figure 1.18, which is comparable to the paratungstate A anion.[81–84]
- Octamolybdate $[Mo_8O_{26}]^{4-}$ is an IPA motif, which was first reported in 1930 by Hubert Britton and William German. Further

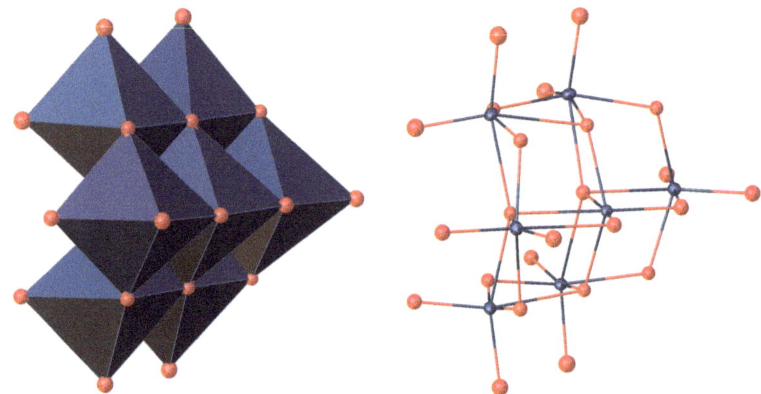

Figure 1.17 The paratungstate (A) structure $[W_7O_{24}]^{6-}$. Polyhedron (left) and atomistic representation (right). Color code: blue – W and red – O. The data were used from the Cambridge Crystallographic Data Centre and Fachinformationszentrum Karlsruhe Access Structures service database (deposition number: 1565298).

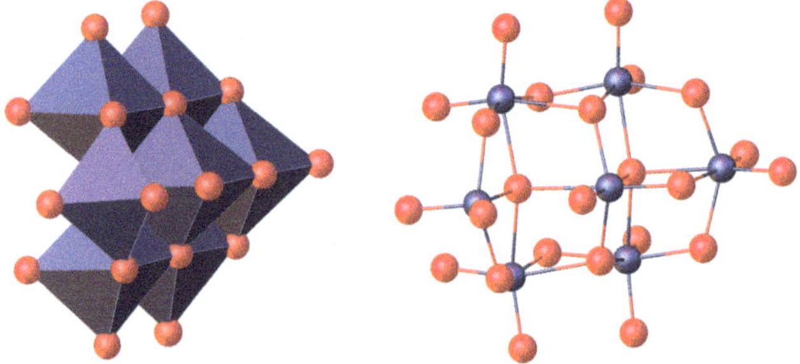

Figure 1.18 The paramolybdate structure $[Mo_7O_{24}]^{6-}$. Polyhedron (left) and atomistic representation (right). Color code: blue – Mo and red – O. The data were used from the Cambridge Crystallographic Data Centre and Fachinformationszentrum Karlsruhe Access Structures service database (deposition number: 1110190).

investigations have been done since 1959. All MoO_6 octahedrons are linked *via* edge sharing. The structural motif has a parallelogram-shaped orientation, with each edge consisting of two octahedrons. It is formed by a simple acidification of aqueous MoO_4^{2-} solutions at pH values around 6 to 5.[85–87] The structure is visualized in Figure 1.19.

- $[Mo_{36}O_{112}(H_2O)_{16}]^{8-}$ is an IPA structural motif, which was reported in 1979 by Irene Paulat-Böschen. This structure is a polymolybdate anion of mainly edge sharing MoO_6 octahedrons.[88,89]
- Molybdenum blue or POM rings are compounds of the types $[Mo_{154}O_{462}H_8(H_2O)_{70}]^{20-}$ or $[Mo_{154}O_{462}H_{14}(H_2O)_{70}]^{14-}$, or $\{Mo_{154}\}$

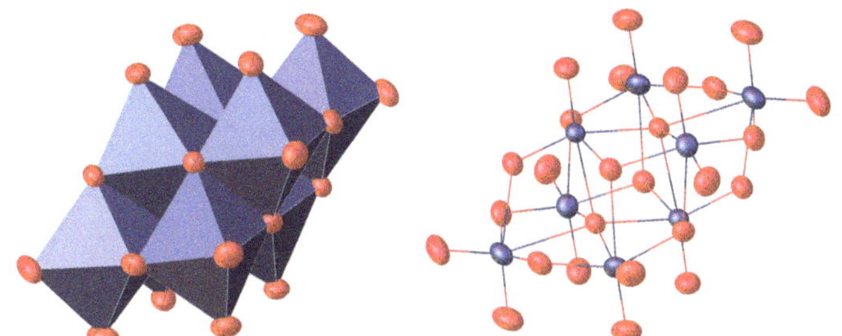

Figure 1.19 The octamolybdate structure $[Mo_8O_{26}]^{4-}$. Polyhedron (left) and atomistic representation (right). Color code: blue – Mo and red – O. The data were used from the Cambridge Crystallographic Data Centre and Fachinformationszentrum Karlsruhe Access Structures service database (deposition number: 2279329).

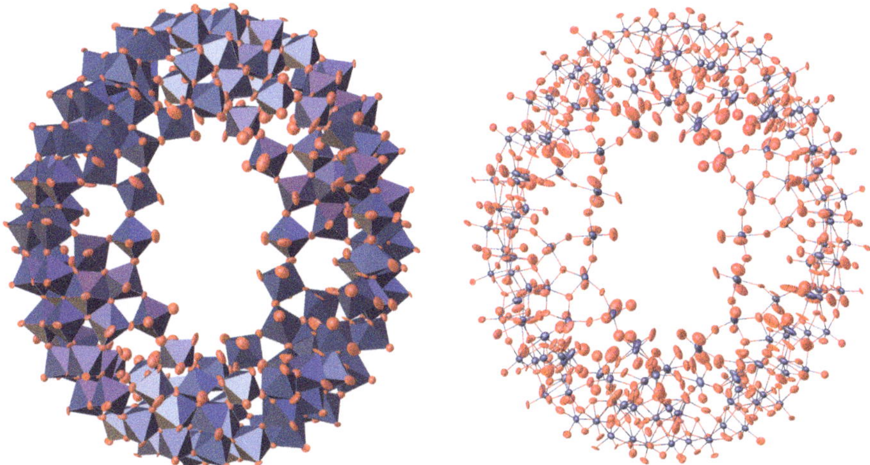

Figure 1.20 The molybdenum blue structure $\{Mo_{154}\}$. Polyhedron (left) and atomistic representation (right). Color code: blue – Mo and red – O. The data were used from the Cambridge Crystallographic Data Centre and Fachinformationszentrum Karlsruhe Access Structures service database (deposition number: 745074).

for short. The so-called molybdenum blue or giant wheel rings $\{Mo_{154}\}$ are gigantic molecular rings with large inner cavities.[90] The first examples were reported by Carl Wilhelm Scheele in 1778, with the first information about their composition coming from Berzelius in 1826.[91,92] These clusters belong to the class of IPA structures.[93] The Mo atoms are present in oxidation states of +5 and +6, with the characteristic blue color caused by the reduced Mo^{V} species. The anions $\left[Mo^{VI}_{126}Mo^{V}_{28}O_{462}H_{14}(H_2O)_{70}\right]^{15-}$ and $\left[Mo^{VI}_{118}Mo^{V}_{28}O_{442}H_{14}(H_2O)_{58}\right]^{22-}$ are known.[94] A typical

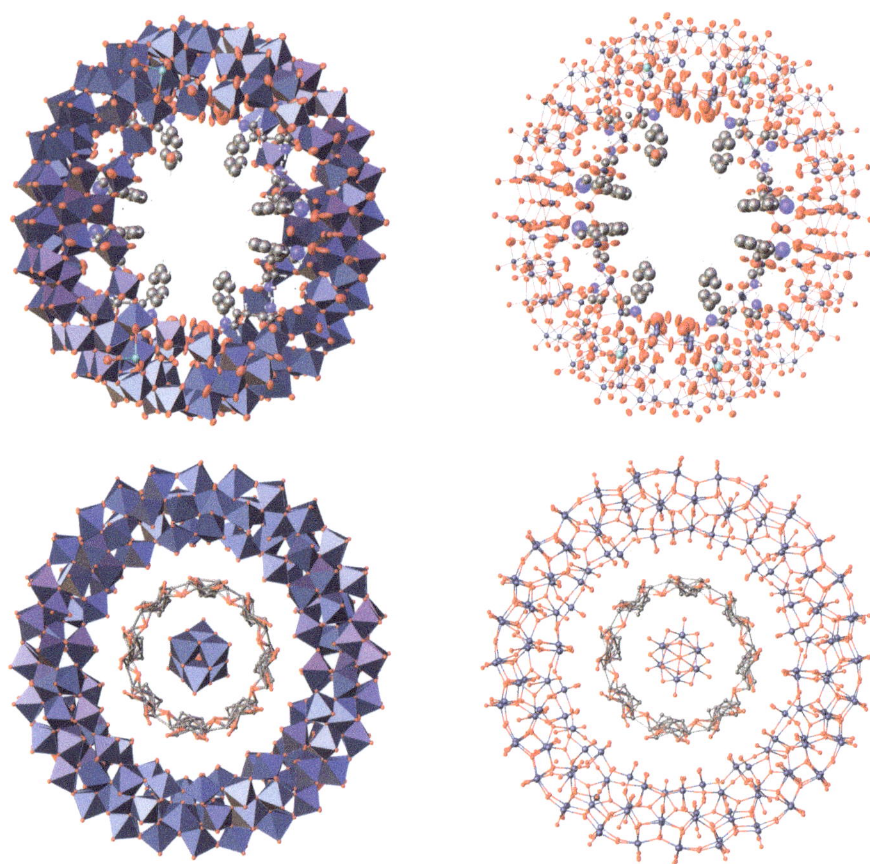

Figure 1.21 Molybdenum blue structures $\{Mo_{154}\}$, modified with organic ligands. Polyhedron (left) and atomistic representation (right). Color code: blue - Mo, red - O, gray - C, light blue - N. The data were used from the Cambridge Crystallographic Data Centre and Fachinformationszentrum Karlsruhe Access Structures service database (deposition numbers: 1853698 - top and 2020274 - bottom).

representation of these clusters is shown in Figure 1.20. Small, organic molecules can coordinate to the Mo atoms in the inner spaces of the rings, as shown in Figure 1.21.[95,96]

Abbreviations

POM	Polyoxometalate
pH	Negative decadic logarithm of the hydrogen ionconcentration
HPA	Heterpolyanion/heteropolyacid

IPA	Isopolyanion/isopolyacid
sc-XRD	Single-crystal X-ray diffraction
DFT	Density functional theory

Acknowledgements

I would like to thank the publisher, The Royal Society of Chemistry, for the opportunity to write this book!

Recommended Reading

Please have a look into the following references for an alternative overview of polyoxometalates:

1. M. T. Pope and A. Müller, *Polyoxometalate Chemistry From Topology via Self-Assembly to Applications*, Kluwer Academic Publishers, New York, Boston, Dordrecht, London, Moscow, 2002.
2. J. J. Borrás-Almenar, E. Coronado, A. Müller and M. Pope, *Polyoxometalate Molecular Science*, Springer Netherlands, Dordrecht, 2003.
3. T. Yamase and M. T. Pope, *Polyoxometalate Chemistry for Nano-Composite Design*, Kluwer Academic Publishers, New York, Boston, Dordrecht, London, Moscow, 2004.
4. E. Coronado and C. J. Gómez-García, *Chem. Rev.*, 1998, **98**, 273–296.
5. D.-L. Long, R. Tsunashima and L. Cronin, *Angew. Chem.*, 2010, **122**, 1780–1803.
6. R. Dehghani, S. Aber and F. Mahdizadeh, *Clean – Soil, Air, Water*, 2018, **46**, 1800413.
7. D. L. Long, E. Burkholder and L. Cronin, *Chem. Soc. Rev.*, 2007, **36**, 105–121.
8. M. T. Pope, M. Sadakane and U. Kortz, *Eur. J. Inorg. Chem.*, 2019, **2019**, 340–342.
9. M. Poper, A. Müller, *Angew. Chem., Int. Ed. Engl.*, 1991, **30**, 34–48.
10. D.-L. Long, R. Tsunashima, L. Cronin, *Angew. Chem., Int. Ed.*, 2010, **49**, 1736–1758.
11. M. Ammam, *J. Mater. Chem., A*, 2013, **1**, 6291–6312.
12. Y. Wei, *Polyoxometalates*, 2022, **1**, 9140014.
13. B. Li and L. Wu, *Polyoxometalates*, 2023, **2**, 9140016.
14. U. Kortz, A. Müller, J. van Slageren, J. Schnack, N. S. Dalal, M. Dressel, *Coord. Chem. Rev.*, 2009, **253**, 2315–2327.
15. A. Müller, E. Beckmann, H. Bögge, M. Schmidtmann and A. Dress, *Angew. Chem., Int. Ed.*, 2002, **41**, 1162–1167.

References

1. A. F. Holleman, E. und N. Wiberg and G. Fischer, *Lehrbuch Der Anorganischen Chemie*, Berlin, New York, 2009.
2. M. T. Pope and A. Müller, *Polyoxometalate Chemistry From Topology via Self-Assembly to Applications*, Kluwer Academic Publishers, New York, Boston, Dordrecht, London, Moscow, 2002.
3. J. J. Borrás-Almenar, E. Coronado, A. Müller and M. Pope, in *Polyoxometalate Molecular Science*, Springer Netherlands, Dordrecht, 2003.
4. T. Yamase and M. T. Pope, in *Polyoxometalate Chemistry for Nano-Composite Design*, Kluwer Academic Publishers, New York, Boston, Dordrecht, London, Moscow, 2004.
5. E. Coronado and C. J. Gómez-García, *Chem. Rev.*, 1998, **98**, 273–296.
6. D.-L. Long, R. Tsunashima and L. Cronin, *Angew. Chem.*, 2010, **122**, 1780–1803.
7. R. Dehghani, S. Aber and F. Mahdizadeh, *Clean: Soil, Air, Water*, 2018, **46**, 1800413.
8. D. L. Long, E. Burkholder and L. Cronin, *Chem. Soc. Rev.*, 2007, **36**, 105–121.
9. J. Albert, J. Mehler, J. Tucher, K. Kastner and C. Streb, *ChemistrySelect*, 2016, **1**, 2889–2894.
10. A. ÖzbekH, *Chem. Pap.*, 2023, **77**, 5663–5669.
11. M. Nyman, T. M. Alam, F. Bonhomme, M. A. Rodriguez, C. S. Frazer and M. E. Welk, *J. Cluster Sci.*, 2006, **17**, 197–219.
12. D. J. Sures, S. K. Sahu, P. I. Molina, A. Navrotsky and M. Nyman, *ChemistrySelect*, 2016, **1**, 1858–1862.
13. J. J. Berzelius, *Ann. Phys.*, 1826, **82**, 369–392.
14. J.-C. G. de Marignac, *Ann. Chim. Phys.*, 1860, **3**, 257–307.
15. A. Rosenheim and O. Liebknecht, *Justus Liebigs Ann. Chem.*, 1899, **308**, 40–67.
16. M. T. Pope, M. Sadakane and U. Kortz, *Eur. J. Inorg. Chem.*, 2019, 340–342.
17. A. Miolati and R. Pizzighelli, *J. Prakt. Chem.*, 1908, **77**, 417–456.
18. L. Pauling, *J. Am. Chem. Soc.*, 1929, **51**, 2868–2880.
19. J. F. Keggin, *Proc. R. Soc. London, Ser. A*, 1934, **144**, 75–100.
20. F. Kehrmann, *Z. Anorg. Chem.*, 1894, **7**, 406–426.
21. A. Rosenheim and J. Jaenicke, *Z. Anorg. Allg. Chem.*, 1917, **101**, 235–275.
22. B. Dawson, *Acta Crystallogr.*, 1953, **6**, 113–126.
23. A. F. Wells, *Structural Inorganic Chemistry*, Oxford, 1984.
24. I.-M. Mbomekalle, Y. W. Lu, B. Keita and L. Nadjo, *Inorg. Chem. Commun.*, 2004, 7, 86–90.
25. C. R. Graham and R. G. Finke, *Inorg. Chem.*, 2008, **47**, 3679–3686.
26. P. Wu, Y. Wang, B. Huang and Z. Xiao, *Nanoscale*, 2021, **13**, 7119–7133.
27. A. A. Mukhacheva, V. V. Volchek, V. V. Yanshole, N. B. Kompankov, A. L. Gushchin, E. Benassi, P. A. Abramov and M. N. Sokolov, *Inorg. Chem.*, 2020, **59**, 2116–2120.
28. A. Bijelic and A. Rompel, *Acc. Chem. Res.*, 2017, **50**, 1441–1448.
29. H. J. Lunk and H. Hartl, *ChemTexts*, 2021, 7, 1–30.
30. J. S. Anderson, *Nature*, 1937, **140**, 1937.
31. H. T. Evans, *J. Am. Chem. Soc.*, 1948, **70**, 1291–1292.
32. I. Lindqvist, O. Hassel, M. Webb and M. Rottenberg, *Acta Chem. Scand.*, 1950, **4**, 1066–1074.
33. O. Linnenberg, A. Kondinski and K. Y. Monakhov, in *Supramolecular Systems Chemistry*, 2017, pp. 39–66.
34. O. Nagano and Y. Sasaki, *Acta Crystallogr.*, 1979, **35**, 2387–2389.
35. X. López, J. J. Carbó, C. Bo and J. M. Poblet, *Chem. Soc. Rev.*, 2012, **41**, 7537.
36. A. J. Bridgeman and G. Cavigliasso, *Faraday Discuss.*, 2003, **124**, 239.

37. F.-Q. Zhang, H.-S. Wu, Y.-Y. Xu, Y.-W. Li and H. Jiao, *J. Mol. Model.*, 2006, **12**, 551–558.
38. G. Mestl, T. Ilkenhans, D. Spielbauer, M. Dieterle, O. Timpe, J. Kröhnert, F. Jentoft, H. Knözinger and R. Schlögl, *Appl. Catal., A*, 2001, **210**, 13–34.
39. A. Patel, N. Narkhede, S. Singh and S. Pathan, *Catal. Rev.: Sci. Eng.*, 2016, **58**, 337–370.
40. Y. Sakai, A. Shinohara, K. Hayashi and K. Nomiya, *Eur. J. Inorg. Chem.*, 2006, 163–171.
41. L. A. Combs-Walker and C. L. Hill, *Inorg. Chem.*, 1991, **30**, 4016–4026.
42. C. Marchal-Roch, E. Ayrault, L. Lisnard, J. Marrot, F.-X. Liu and F. Sécheresse, *J. Cluster Sci.*, 2006, **17**, 283–290.
43. L. C. W. Baker, G. A. Gallagher and T. P. McCutcheon, *J. Am. Chem. Soc.*, 1953, **75**, 2493–2495.
44. D. D. Dexter and J. V. Silverton, *J. Am. Chem. Soc.*, 1968, **90**, 3589–3590.
45. R. D. Hall, *J. Am. Chem. Soc.*, 1907, **29**, 692–714.
46. L. C. W. Baker and T. J. R. Weakley, *J. Inorg. Nucl. Chem.*, 1966, **28**, 447–454.
47. J. L. T. Waugh, D. P. Shoemaker and L. Pauling, *Acta Crystallogr.*, 1954, **7**, 438–441.
48. R. Allmann and H. D'Amour, *Z. Kristallogr. – Cryst. Mater.*, 1975, **141**, 342–353.
49. L. Y. Feng, Y. H. Wang, Y. J. Qi, C. W. Hu, Y. Xu and E. B. Wang, *J. Mol. Struct.*, 2003, **645**, 231–234.
50. R. Strandberg, L. Niinistö, J. Møller, G. Schroll, K. Leander and C.-G. Swahn, *Acta Chem. Scand.*, 1973, **27**, 1004–1018.
51. P. Pérez-Romo, C. Potvin, J.-M. Manoli and G. Djéga-Mariadassou, *J. Catal.*, 2002, **205**, 191–198.
52. R. Thouvenot, A. Teze, R. Contant and G. Herve, *Inorg. Chem.*, 1988, **27**, 524–529.
53. F. Kehrmann, *Z. Anorg. Chem.*, 1892, **1**, 423–441.
54. J.-C. Raabe, F. Jameel, M. Stein, J. Albert and M. J. Poller, *Dalton Trans.*, 2024, **53**, 454–466.
55. I. Creaser, M. C. Heckel, R. J. Neitz and M. T. Pope, *Inorg. Chem.*, 1993, **32**, 1573–1578.
56. C. Dey, *Coord. Chem. Rev.*, 2024, **510**, 215847.
57. R. Kato, A. Kobayashi and Y. Sasaki, *J. Am. Chem. Soc.*, 1980, **102**, 6571–6572.
58. O. Glemser, W. Holznagel, W. Höltje and E. Schwarzmann, *Z. Naturforsch., B: J. Chem. Sci.*, 1965, **20**, 725–746.
59. E. Birkholz, J. Fuchs, W. Schiller and H.-P. Stock, *Z. Naturforsch., B: J. Chem. Sci.*, 1971, **26**, 365–366.
60. S. C. Termes and M. T. Pope, *Inorg. Chem.*, 1978, **17**, 500–501.
61. T. Yamase, N. Takabayashi and M. Kaji, *J. Chem. Soc., Dalton Trans.*, 1984, 793.
62. A. Chemseddine, C. Sanchez, J. Livage, J. P. Launay and M. Fournier, *Inorg. Chem.*, 1984, **23**, 2609–2613.
63. T. Yamase, T. Kobayashi, M. Sugeta and H. Naruke, *J. Phys. Chem. A*, 1997, **101**, 5046–5053.
64. T. Yamase, H. Naruke and Y. Sasaki, *J. Chem. Soc., Dalton Trans.*, 1990, 1687.
65. T. J. R. Weakley, H. T. Evans, J. S. Showell, G. F. Tourné and C. M. Tourné, *J. Chem. Soc., Chem. Commun.*, 1973, 139–140.
66. D.-F. Shen, S. Li, H. Liu, W. Jiang, Q. Zhang and G.-G. Gao, *J. Mater. Chem. C*, 2015, **3**, 12090–12097.
67. J. Iball, J. N. Low and T. J. R. Weakley, *J. Chem. Soc., Dalton Trans.*, 1974, **7**, 2021–2024.
68. S. Lis, *Acta Phys. Pol., A*, 1996, **90**, 275–283.
69. T. Ozeki, M. Takahashi and T. Yamase, *Acta Crystallogr., Sect. C: Cryst. Struct. Commun.*, 1992, **48**, 1370–1374.
70. E. Rakovský, L. Zúrková and J. Marek, *Cryst. Res. Technol.*, 2001, **36**, 339–344.
71. H. T. Evans, *Inorg. Chem.*, 1966, **5**, 967–977.

72. A. G. Swallow, F. R. Ahmed and W. H. Barnes, *Acta Crystallogr.*, 1966, **21**, 397–405.
73. V. F. Odyakov and E. G. Zhizhina, *Russ. J. Inorg. Chem.*, 2009, **54**, 361–367.
74. V. F. Odyakov, E. G. Zhizhina and R. I. Maksimovskaya, *Appl. Catal., A*, 2008, **342**, 126–130.
75. H. T. Evans and O. W. Rollins, *Acta Crystallogr., Sect. B*, 1976, **32**, 1565–1567.
76. J. Chojnacka, E. Hodorowicz and S. Sagnowski, *J. Inorg. Nucl. Chem.*, 1976, **38**, 1811–1813.
77. R. I. Maksimovskaya and K. G. Burtseva, *Polyhedron*, 1985, **4**, 1559–1562.
78. Q. Gao, D.-H. Hu, M.-H. Duan and D.-H. Li, *J. Mol. Struct.*, 2019, **1184**, 400–404.
79. J. Martín-Caballero, B. Artetxe, S. Reinoso, L. San Felices, O. Castillo, G. Beobide, J. L. Vilas and J. M. Gutiérrez-Zorrilla, *Chem. – Eur. J.*, 2017, **23**, 14962–14974.
80. Y. Guo, D. Li, C. Hu, Y. Wang, E. Wang, Y. Zhou and S. Feng, *Appl. Catal., B*, 2001, **30**, 337–349.
81. A. Rosenheim, A. Wolff, E. Koch and N. Siao, *Z. Anorg. Allg. Chem.*, 1930, **193**, 47–63.
82. A. Rosenheim, M. Hakki and O. Krause, *Z. Anorg. Allg. Chem.*, 1932, **209**, 175–203.
83. A. R. Middleton, *J. Chem. Educ.*, 1933, **10**, 726.
84. J. H. Sturdivant, *J. Am. Chem. Soc.*, 1937, **59**, 630–631.
85. H. T. S. Britton and W. L. German, *J. Chem. Soc.*, 1930, 2154–2166.
86. P. Cannon, *J. Inorg. Nucl. Chem.*, 1959, **9**, 252–266.
87. H. E. Garrett and A. J. Walker, *Analyst*, 1964, **89**, 642.
88. I. Paulat-Böschen, *J. Chem. Soc., Chem. Commun.*, 1979, 780–782.
89. B. Krebs and I. Paulat-Böschen, *Acta Crystallogr., Sect. B*, 1982, **38**, 1710–1718.
90. A. Müller, E. Krickemeyer, J. Meyer, H. Bögge, F. Peters, W. Plass, E. Diemann, S. Dillinger, F. Nonnenbruch, M. Randerath and C. Menke, *Angew. Chem., Int. Ed. Engl.*, 1995, **34**, 2122–2124.
91. A. Müller, J. Meyer, E. Krickemeyer and E. Diemann, *Angew. Chem., Int. Ed. Engl.*, 1996, **35**, 1206–1208.
92. A. Müller and C. Serain, *Acc. Chem. Res.*, 2000, **33**, 2–10.
93. A. Müller, D. Fenske and P. Kögerler, *Curr. Opin. Solid State Mater. Sci.*, 1999, **4**, 141–153.
94. A. Müller, S. K. Das, V. P. Fedin, E. Krickemeyer, C. Beugholt, H. Bögge, M. Schmidtmann and B. Hauptfleisch, *Z. Anorg. Allg. Chem.*, 1999, **625**, 1187–1192.
95. W. Xuan, R. Pow, N. Watfa, Q. Zheng, A. J. Surman, D.-L. Long and L. Cronin, *J. Am. Chem. Soc.*, 2019, **141**, 1242–1250.
96. C. Falaise, S. Khlifi, P. Bauduin, P. Schmid, W. Shepard, A. A. Ivanov, M. N. Sokolov, M. A. Shestopalov, P. A. Abramov, S. Cordier, J. Marrot, M. Haouas and E. Cadot, *Angew. Chem., Int. Ed.*, 2021, **60**, 14146–14153.

2 Tendencies of the Elements From the Periodic Table to Form Oxo Compounds

2.1 General Tendencies

The thermodynamic driving force for the oxo compound formation lies in the so-called oxophilicity of an element. Oxophilicity is the tendency of an element to form a covalent bond with oxygen. Oxygen is the element with the second strongest electronegativity. This means that high electronegativity differences are achieved in a covalent bond between an element and oxygen. The element with an even higher electronegativity is the halogen fluorine F. Therefore, the tendency of elements to form a covalent bond with F is even higher than with oxygen. This is also the reason why oxides are soluble in hydrofluoric acid (HF) solutions and form fluorides. One example is the decomposition of silicon dioxide (SiO_2) with HF to $H_2[SiF_6]$.[1–3] Eqn (2.1) shows the reaction equation of the degradation of SiO_2 by F^- (element with the highest electronegativity).

$$SiO_2 + 6HF \rightarrow H_2[SiF_6] + 2H_2O \tag{2.1}$$

RSC Foundations No. 3
Polyoxometalate Chemistry
By Jan-Christian Raabe

Published by the Royal Society of Chemistry, www.rsc.org

2.2 Oxo Compounds of the Main-group Elements

The simplest oxo compounds from the periodic table are formed with hydrogen and alkali metals. For hydrogen, water H_2O and hydrogen peroxide H_2O_2 are known. Alkali and alkaline earth metals form hydroxides like MOH and $M(OH)_2$.[1]

A summary of the most important oxo compounds formed by main-group elements is shown in Table 2.1. For the noble gas xenon, oxo compounds are known, with xenon being the most reactive noble gas. Oxygen forms oxo compounds with itself, the dioxygen O_2 and ozone O_3 molecules. Oxo compounds for main-group elements in defined oxidation states are shown in Figure 2.1 for carbon, phosphorus, sulfur, the halogens and the noble gas xenon.[1]

- Carbon: formaldehyde ($H_2C^{0}O$), formic acid ($HC^{+II}OOH$), carbonic acid ($C^{+IV}O(OH)_2$), carbon dioxide ($C^{+IV}O_2$). Salts: formiate ($HC^{+II}OO^-$) and carbonate ($C^{+IV}O_3{}^{2-}$).[1]

Table 2.1 Oxo compounds formed by the main-group elements from the periodic table.[1]

Group 13	Group 14	Group 15	Group 16	Group 17	Group 18
$B(OH)_3$, $BO_3{}^{3-}$	H_2CO, HCOOH, $CO(OH)_2$, CO_2, Cl_2CO	HNO_3	O_2, O_3	HFO	—
AlOCl, $Al(OH)_3$, $[Al(OH)_4]^-$	ClSiO, $Si(OH)_4$	P_4O_{10}, P_4O_6, H_3PO_4, H_3PO_3, H_3PO_2, $POCl_3$	SO_x ($x=1$ to 3), H_2SO_4, H_2SO_3, H_2SO_2	HClO, $HClO_2$, $HClO_3$, $HClO_4$	—
GaOCl, $Ga(OH)_3$	Cl_2GeO, $[GeO_4]^{4-}$	H_3AsO_4, H_3AsO_3, H_3AsO_2, $AsOCl_3$	SeO_x ($x=1$ to 3), H_2SeO_4, H_2SeO_3	HBrO, $HBrO_2$, $HBrO_3$, $HBrO_4$	—
InOCl, $In(OH)_3$	Cl_2SnO, $Sn(OH)_2$	Sb_2O_3 $Sb(OH)_3$, $Sb(OH)_5$, $K[Sb(OH)_6]$	TeO_x ($x=1$ to 3), $Te(OH)_6$, H_2TeO_3, H_2TeO_2	HIO, HIO_2, HIO_3, HIO_4	XeO_4, XeO_3, XeO_2, $XeOF_4$, $XeOF_2$, XeO_2F_2, XeO_2F_4, XeO_3F_2
Tl(OH)	Cl_2PbO, $Pb(OH)_2$	$Bi(OH)_3$, H_3BiO_4	—	—	—

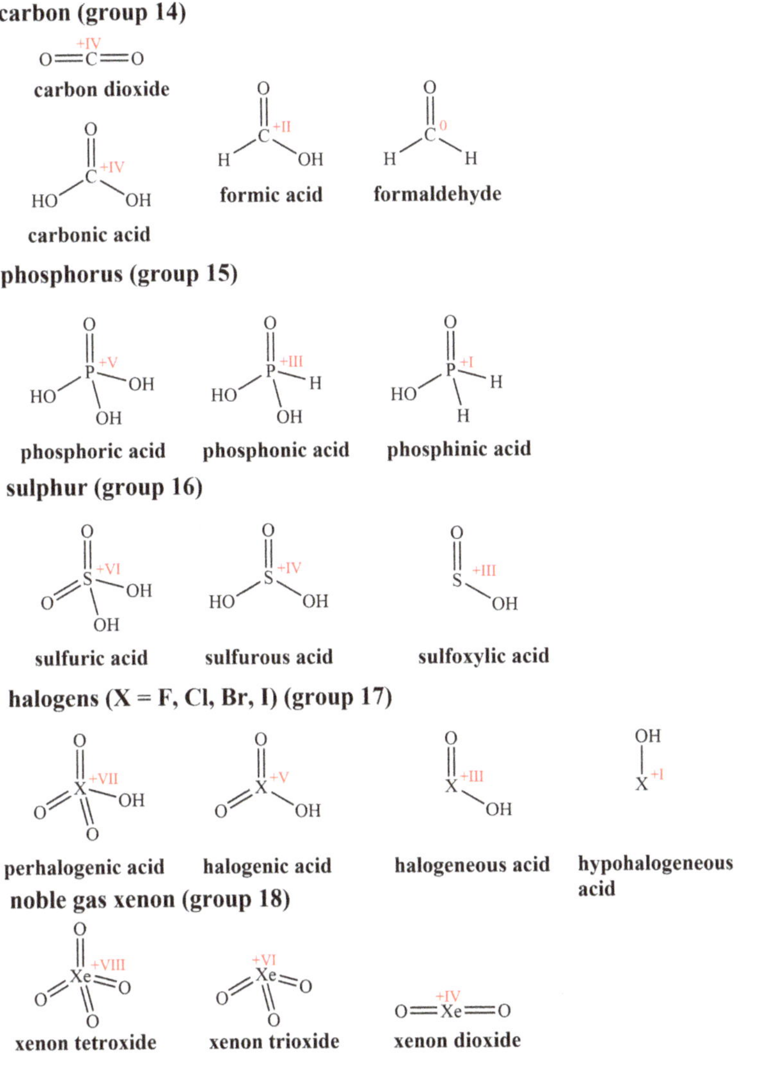

Figure 2.1 Representative oxo compounds from group 14 to 18 elements in their defined oxidation states.[1]

- Phosphorus: phosphoric acid ($H_3P^{+V}O_4$), phosphonic acid ($H_3P^{+III}O_3$), phosphinic acid ($H_3P^{+I}O_2$). Salts: phosphate ($P^{+V}O_4{}^{3-}$), phosphonate ($P^{+III}O_3{}^{2-}$), phosphinate ($P^{+I}O_2{}^{-}$).[1]
- Sulfur: sulfuric acid ($H_2S^{+VI}O_4$), sulfurous acid ($H_2S^{+IV}O_3$), sulfoxylic acid ($H_2S^{+III}O_2$). Salts: sulfate ($S^{+VI}O_4{}^{2-}$), sulfite ($S^{+IV}O_3{}^{2-}$), hyposulfite ($S^{+III}O_2{}^{2-}$).[1]
- Halogens: perhalogenic acid ($HX^{+VII}O_4$), halogenic acid ($HX^{+V}O_3$), halogeneous acid ($HX^{+III}O_2$), hypohalogeneous acid

($HX^{+I}O$). Salts: perhalogenate ($X^{+VII}O_4^-$), halogenate ($X^{+V}O_3^-$), halogenite ($X^{+III}O_2^-$), hypohalogenite ($X^{+I}O^-$).[1]
- Xenon: xenon tetroxide ($Xe^{+VIII}O_4$), xenon trioxide $Xe^{+VI}O_3$, xenon dioxide $Xe^{+IV}O_2$.[1]

All mineral acids form defined salts, which are derived from the respective element name. For example: sodium perhalogenate $NaXO_4$, sodium halogenate $NaXO_3$, sodium halogenite $NaXO_2$ and sodium hypohalogenite NaXO.[1]

The phosphorus oxides can also be interpretated as oxo compounds, because the oxides are molecular compounds with a defined number of atoms. The noble gases of period 5 and above deviate from their noble behavior and form chemical bonds with elements of high electronegativity, like O or F.[1]

2.3 Oxo Compounds of the Transition Elements

For main-group oxo compounds, the s and p orbitals are involved in bond formation. For transition element oxo compounds, d orbitals are also relevant for a detailed discussion.

An octahedrally coordinated oxo compound is discussed as an example. Here, the six oxo ligands come closer to the $d_{x^2-y^2}$ orbitals lying on the x/y axes and the d_{z^2} orbital on the z axis than to the d_{xy}, d_{xz} and d_{yz} orbitals lying between the axes. This means that the $d_{x^2-y^2}$ and d_{z^2} orbitals are energetically raised (e_g* orbitals) and the d_{xy}, d_{xz} and d_{yz} orbitals are lowered (t_{2g}* orbitals). The energy difference between the e_g and t_{2g} levels corresponds to the splitting of the octahedral ligand field Δ_{oct}. If the oxo ligands are now considered, the s and p orbitals are relevant for bond formation with the metal. Each of the oxo ligands has a set of molecular orbitals (MOs) a_{1g}/a_{1g}*, t_{1u}/t_{1u}* and e_g. However, the oxo ligands are π-donor ligands, meaning that the occupied ligand p orbitals can transfer electrons to the metal, with participation of the unoccupied d_{xy}, d_{xz} and d_{yz} orbitals. A set of three MOs t_{2g} is formed. The situation is illustrated in Figure 2.2.[3]

In general, all MOs marked with a * represent antibonding MOs. The t_{2g}* and e_g* MOs (with the energy difference of Δ_{oct}) are also antibonding MOs. Considering an oxo compound with octahedral coordination geometry and a transition element in its highest oxidation state (d^0 configuration), like Mo(VI), W(VI), V(V), Nb(V) or Ta(V), the six π-donor ligands provide 18 electrons (12σ and 6π electrons), which occupy the binding MOs a_{1g}, t_{1u}, e_g and t_{2g} completely. This

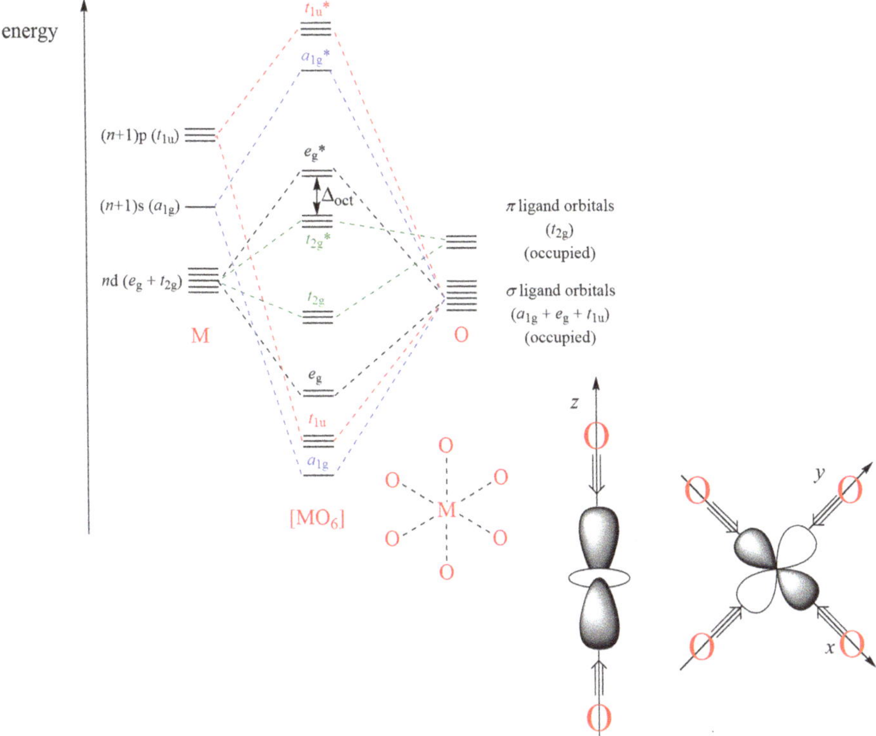

Figure 2.2 MO diagram of an octahedral metal complex for π-donor ligands.

means that for a d^0 element (in which no d electron is available) the antibonding MOs are completely unoccupied, also t_{2g}* and e_g*. So, the metal oxygen bond is maximally stabilized for d^0 transition elements. However, as soon as transition elements with a d^n configuration are considered, where n is not equal to zero, the n valence electrons have to occupy the antibonding MOs, especially t_{2g}* and/or e_g*. With increasing occupancy of the antibonding MOs, the metal oxygen bond is increasingly weakened. This is the reason why the most stable oxo compounds are formed with transition elements in their highest or in very high oxidation states.[3] An example is the permanganate anion MnO_4^-, with Mn(VII) in its highest oxidation state of +7. A summary of the most popular oxo compounds formed by transition elements is shown in Table 2.2.[1,4,5]

The most popular oxo compounds are formed by V (group 5), chromium Cr, Mo, W (group 6) and Mn (group 7). Here the metavanadate anion (VO_3^-), pervanadyl cation (VO_2^+), vanadyl cation (VO^{2+}), chromate anion (CrO_4^{2-}), molybdate anion (MoO_4^{2-}), tungstate anion (WO_4^{2-}), permanganate anion (MnO_4^-), pertechnetate

Table 2.2 Oxo compounds formed by transition elements of the periodic table.[1,4,5]

Group 3	Group 4	Group 5	Group 6	Group 7	Group 8
SOCl	$TiOCl_2$	VO_3^-, VO_2^+, VO^{2+}	CrO_4^{2-}	MnO_4^-	FeOCl
YOCl	$ZrOCl_2$	—	MoO_4^{2-}	TcO_4^-	RuOCl
—	$HfOCl_2$	—	WO_4^{2-}	ReO_4^-	OsOCl

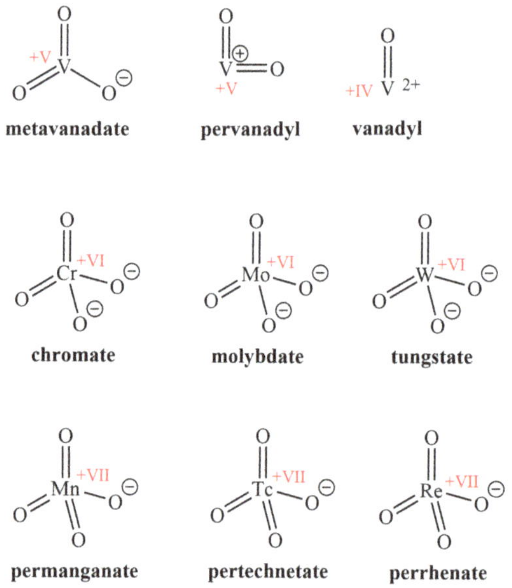

Figure 2.3 Well-known oxo compounds formed by transition elements from group 5 to 7 of the periodic table.

anion (TcO_4^-) and perrhenate anion (ReO_4^-) are known. Structures and oxidation states of those elements in their oxo compounds are shown in Figure 2.3.[1]

The cation VO_2^+ is found in a bent arrangement. Formally, both oxo ligands can be regarded as "*cis*" to each other. This geometry is a consequence of the *trans* effect and/or *trans* influence in the ground state, as shown in Figure 2.4. A strong *trans* effect is only observed with σ donor and π acceptor ligands, which form a σ bond with the d_{z^2} orbital in the first step. Those ligands reduce the electron density in the d_{z^2} orbital due to the π backdonation and weaken the bond to the *trans* orientated (opposite) ligand, which can only overlap with the d_{z^2} orbital. A coordination of the ligand in *trans* position is unfavored, meaning that the binding affinity is low. The binding affinity of *cis* oriented ligands is unchanged, since the electron density in the $d_{x^2-y^2}$ orbital is unchanged. This means that the electron density in the *cis*

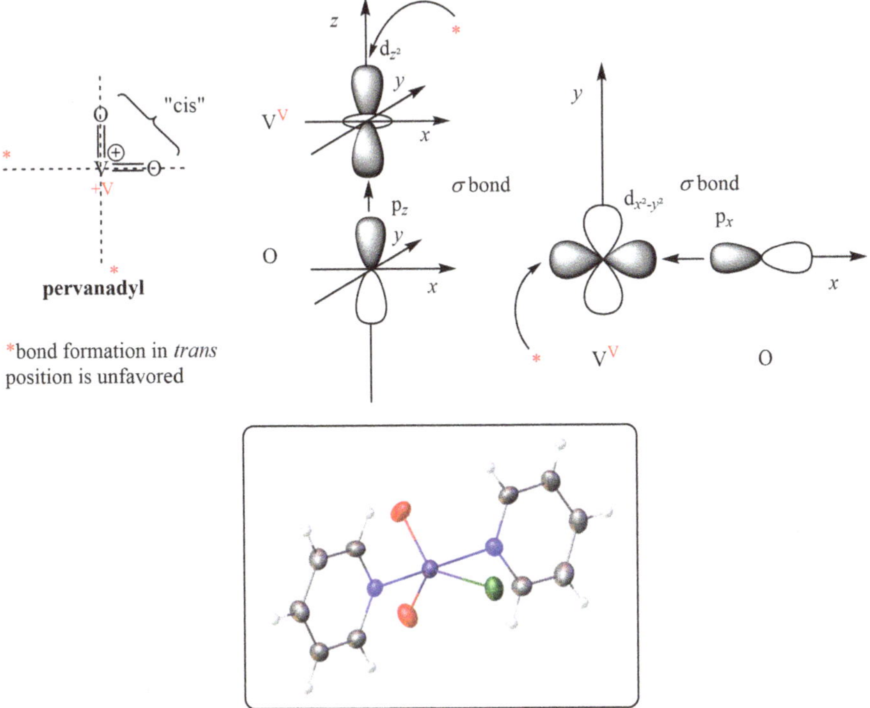

Figure 2.4 Explanation of the bent arrangement of the VO_2^+ cation. Only σ bond formation (with the d_{z^2} and $d_{x^2-y^2}$ orbital) of both oxo ligands is discussed. In the box, the solid-state structure of the complex $[(VO_2)Clpy_2]$ is shown, with a significant elongation of N–V and Cl–V bonds in comparison to the V=O bonds. Color code: blue - V, sky blue - N, red - O, green - Cl, gray - C and white - H. The data were used from the Cambridge Crystallographic Data Centre and Fachinformationszentrum Karlsruhe Access Structures service database (deposition number: 687017).[6]

oriented $d_{x^2-y^2}$ orbital is higher compared to the d_{z^2} orbital. Another ligand preferentially binds to a *cis* position and overlaps with the $d_{x^2-y^2}$ orbital.[7]

However, oxo ligands are π donor ligands, as discussed above. Those ligands form a σ and π donor bond. The formation of two donor bonds is preferred for V in its oxidation state of +5, and generally for d^0 elements, as the metal itself has no electrons available and is therefore dependent on an electron transfer from the ligands. For the σ donor bond, electrons are transferred from the σ orbital of the oxo ligand to the d_{z^2} orbital of the metal(v). To form a π donor bond, occupied π orbitals of the oxo ligand transfer electron density to the unoccupied d_{xy} orbital of the metal(v). This means that electron density is transferred to the metal by two types of bond, the σ and π bond. Due to the strong electronegativity of oxygen, an opposite effect is observed, which attracts the electron density to oxygen. These opposing effects cause the metal(v) and oxygen to compete for electron density, weakening the binding affinity of a new oxo ligand in *trans* position. If a second oxo ligand should be bonded to V in *trans* position to the first ligand, the second ligand can only form its σ bond with the d_{z^2} orbital again. The binding affinity therefore is lower than to the $d_{x^2-y^2}$ orbital in *cis* position. Therefore, the second ligand prefers to bind in *cis* position and forms a σ bond with the $d_{x^2-y^2}$ orbital. A π donation bond can be formed by the second oxo ligand with the d_{yz} orbital of V, forming a double bond to V (V=O).[7]

In Figure 2.4 (bottom) the solid-state structure of a chlorido/pyridine (py) complex of the VO_2^+ cation $[(VO_2)Clpy_2]$ is shown, which is coordinated by one chlorido and two py ligands. The V=O bond length is on average 1.619 Å, the Cl–V bond 2.326 Å and the N–V bond 2.139 Å. Bond lengths of both Cl–V and N–V are significantly elongated due to the higher *trans* influence of the oxo ligands in the ground state. A typical bond angle of O–V–O is 110.4° (ref. 6).

2.3.1 Oxo Compounds of Chromium

The condensation behavior (the so-called oligomerization) of oxo compounds formed by transition elements was first analyzed using chromate (CrO_4^{2-}) as an example. CrO_4^{2-} is obtained by the oxidation of chromium(III) oxide (Cr_2O_3). Under acidic conditions CrO_4^{2-} is dimerized to dichromate $Cr_2O_7^{2-}$.[1] Eqn (2.2) shows the formation of $[CrO_4]^{2-}$ and the acid-induced dimerization to $[Cr_2O_7]^{2-}$:

$$Cr_2O_3 + 1.5O_2 \xrightarrow[-2CO_2]{+2Na_2CO_3} 2Na_2CrO_4 \xrightarrow[-Na_2SO_4]{+H_2SO_4} Na_2Cr_2O_7 \quad (2.2)$$

The reaction is noticeable by a color change from yellow (CrO_4^{2-}) to orange ($Cr_2O_7^{2-}$). It occurs mechanistically *via* the protonation of CrO_4^{2-} to $HCrO_4^-$ (hydrogen chromate), which dimerizes to $Cr_2O_7^{2-}$ at room temperature with water splitting, with an equilibrium constant of $K = 10^{2.2}$.[1] Eqn (2.3) shows the reaction mechanism of the $[CrO_4]^{2-}$ dimerization under acidic conditions:

$$\begin{aligned} 4Na^+ + 2CrO_4^{2-} + 2H^+ + SO_4^{2-} &\rightleftharpoons 2Na^+ + 2HCrO_4^- + 2Na^+ + SO_4^{2-} \\ &\rightleftharpoons 2Na^+ + Cr_2O_7^{2-} + 2Na^+ + SO_4^{2-} + H_2O \end{aligned} \quad (2.3)$$

Every CrO_4^{2-} solution therefore also contains $Cr_2O_7^{2-}$ ions and *vice versa*. The equilibrium can be shifted by changing the pH value (pH > 8 only CrO_4^{2-}, pH 2–6 $HCrO_4^-$ and $Cr_2O_7^{2-}$ in equilibrium and pH < 1 only $Cr_2O_7^{2-}$). At even more acidic conditions and higher concentrations of the mineral acids, tri- and tetrachromates ($Cr_3O_{10}^{2-}$ and $Cr_4O_{13}^{2-}$) are formed first, up to the polychromates $[Cr_nO_{3n+1}]^{2-}$, whereby the color of the solution changes to red.

2.3.2 Oxo Compounds of Molybdenum and Tungsten

The most popular oxo compounds of molybdenum and tungsten are MoO_4^{2-} and WO_4^{2-}. During acidification of aqueous solutions of those anions (below pH 7) polymolybdates and -tungstates are formed: hepta-, octa- and oligomolybdate (36 Mo atoms) or tetra-, deca- and dodecatungstate.[1] At pH values around 5, the para-dodecatungstate B anion $[W_{12}O_{40}]^{8-}$ is formed, also in its protonated form $[H_2W_{12}O_{40}]^{6-}$.[1,8] Eqn (2.4) shows the acid-induced/pH dependent oligomerization of $[Mo/WO_4]^{2-}$ in aqueous media:

$$7[MoO_4]^{2-} \underset{-/+4H_2O}{\overset{+/-8H^+}{\rightleftharpoons}} [Mo_7O_{24}]^{6-} \underset{-/+2H_2O}{\overset{+/-MoO_4^{2-},+/-4H^+}{\rightleftharpoons}} [Mo_8O_{26}]^{4-} \underset{-/+10H_2O}{\overset{+/-28MoO_4^{2-},+/-52H^+}{\rightleftharpoons}} [Mo_{36}O_{112}(H_2O)_{16}]^{8-}$$

$$7[WO_4]^{2-} \underset{-/+4H_2O}{\overset{+/-8H^+}{\rightleftharpoons}} [W_7O_{24}]^{6-} \underset{-/+4H_2O}{\overset{+/-3MoO_4^{2-},+/-8H^+}{\rightleftharpoons}} [W_{10}O_{32}]^{4-} \overset{+/-2WO_4^{2-}}{\rightleftharpoons} [W_{12}O_{40}]^{8-} \overset{+/-2H^+}{\rightleftharpoons} [H_2W_{12}O_{40}]^{8-} \quad (2.4)$$

At more acidic pH values, the oxo compounds are directly converted to the metal trioxides MO_3.

2.3.3 Oxo Compounds of Vanadium

A very important vanadium-containing oxo compound is sodium metavanadate $NaVO_3$ (Figure 2.5). This compound is very soluble in warm water. During dissolution in water, the pH value of the aqueous, colorless solution increases to 8.6. With the use of ^{51}V NMR spectroscopy, it was shown that the VO_3^- anion is oligomerized to the tetramer $[V_4O_{12}]^{4-}$ and the pentamer $[V_5O_{15}]^{5-}$, which were identified as intense signals at ∼−575 and ∼−585 ppm. Furthermore, the dimeric, protonated species $[HV_2O_7]^{3-}$ and the orthovanadate $[HVO_4]^{2-}$ were also identified with very low intensity signals at ∼−565 and ∼−540 ppm. After the pH value was acidified to 5, six vanadium species were found: the decavanadate species $[V_{10}O_{28}]^{6-}$ with three

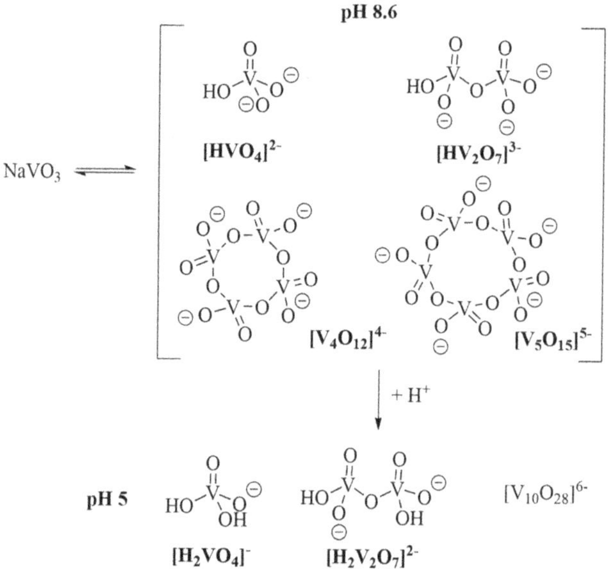

Figure 2.5 Different vanadate species found by dissolving metavanadate in aqueous solution at two pH values.

intense peaks at chemical shifts of ~-425, ~-500 and ~-520 ppm; the double-protonated orthovanadate $[H_2VO_4]^-$ at a shift of ~-560 ppm; the double-protonated, dimeric species $[H_2V_2O_7]^{2-}$ at ~-572 ppm; and again the tetrameric species $[V_4O_{12}]^{4-}$ at ~-575 ppm. During the acidification, the color of the solution turns from colorless to intense red. All the different vanadium species are shown in Figure 2.5. Here, it is shown that the tetra- and the pentameric species are cyclic inorganic compounds.[9,10]

2.3.4 Oxo Compounds of Niobium and Tantalum

In contrast to vanadate chemistry, no ions of the type MO_4^{3-} are known for Nb and Ta. Unlike V_2O_5, the oxides Nb_2O_5 and Ta_2O_5 are not attacked by mineral acids (except HF). However, oligometalates of Nb and Ta can be obtained by basic melt synthesis of the oxides with hydroxides (such as potassium hydroxide KOH), forming the hexametalates (potassium hexaniobate/hexatantalate) $K_8[Nb_6O_{19}]$/$K_8[Ta_6O_{19}]$ of the Lindqvist-type structure. Those POMs can only be formed in basic, aqueous solutions at pH values above 11. This is a significant difference compared to the chemistry of vanadates, as oligovanadates are only formed in acidic pH media (below 8.6) and rearrange to VO_4^{3-} and VO_3^- at basic pH values.[1,11–13] Eqn (2.5) shows the basic degradation of M_2O_5 (M = Nb,Ta) to hexametalates and acid induced oxide formation:

$$\begin{aligned} 3M_2O_5 + 8KOH &\rightarrow K_8[M_6O_{19}] + 4H_2O \\ K_8[M_6O_{19}] + 8HCl &\rightarrow 3M_2O_5 + 8KCl + 4H_2O \end{aligned} \quad (2.5)$$

Below pH values of 7 (Nb) and 10 (Ta), the oxides M_2O_5 precipitate out. The anion $[Nb_6O_{19}]^{8-}$ is protonated at pH values below 11 and forms $H[Nb_6O_{19}]^{7-}$ first and then $H_2[Nb_6O_{19}]^{6-}$. The compound $H_2[Nb_6O_{19}]^{6-}$ precipitates to Nb_2O_5 below pH 7.[1] Eqn (2.6) shows the acid induced decomposition of $[Nb_6O_{19}]^{8-}$ to Nb_2O_5 *via* a stepwise protonation of the $[Nb_6O_{19}]^{8-}$ anion:

$$K_8[Nb_6O_{19}] \xrightarrow[-KCl]{+HCl} K_7H[Nb_6O_{19}] \xrightarrow[-KCl]{+HCl} K_6H_2[Nb_6O_{19}] \xrightarrow[\substack{-6KCl \\ -4H_2O}]{+6HCl} 3Nb_2O_5 \quad (2.6)$$

Abbreviations

POM	Polyoxometalate
pH	Negative decadic logarithm of the hydrogen ion concentration
MO	Molecular orbital
ppm	Parts per million
Å	Ångström (1 Å = 10^{-10} m)
py	Pyridine

Acknowledgements

I would like to thank the publisher, the Royal Society of Chemistry, for the opportunity to write this book!

Recommended Reading

Please have a look into the following references for a better understanding of the concepts of electronegativity, main-group/transition elements and oxophilicity. Have a look at ref. 3 in order to learn how molecular orbital schemes are created. Further effects like the *trans*-influence are also discussed in ref. 3. Note: ref. 1 is only available in German. English version of (2) ISBN-10: 0-470-01864-X

(Hardcover)/ISBN-10: 0-470-01865-8 (Paperbook)/ISBN-13: 978-0-470-05726-1 (E-Book) and (3): ISBN-10: 1-292-13414-3

1. A. F. Holleman, E. und Nils Wiberg and G. Fischer, *Lehrbuch Der Anorganischen Chemie*, Berlin, New York, 2009.
2. U. Müller, *Anorganische Strukturchemie*, Vieweg+Teubner, Wiesbaden, 2008.
3. C. E. Housecroft and A. G. Sharpe, *Anorganische Chemie*, Pearson Studium, München, 2006.

References

1. A. F. Holleman, E. und N. Wiberg and G. Fischer, *Lehrbuch Der Anorganischen Chemie*, Berlin, New York, 2009.
2. U. Müller, *Anorganische Strukturchemie*, Vieweg+Teubner, Wiesbaden, 2008.
3. C. E. Housecroft and A. G. Sharpe, *Anorganische Chemie*, Pearson Studium, München, 2006.
4. H. G. Brittain and G. Meyer, *J. Less-Common Met.*, 1986, **126**, 175–179.
5. W. Levason, J. S. Ogden, A. J. Rest and J. W. Turff, *J. Chem. Soc., Dalton Trans.*, 1982, 1877–1878.
6. M. F. Davis, M. Jura, A. Leung, W. Levason, B. Littlefield, G. Reid and M. Webster, *Dalton Trans.*, 2008, **687015**, 6265.
7. B. Weber, *Koordinationschemie*, Springer, Berlin, Heidelberg, 2014.
8. J.-C. Raabe, F. Jameel, M. Stein, J. Albert and M. J. Poller, *Dalton Trans.*, 2024, **53**, 454–466.
9. J.-C. Raabe, T. Esser, M. J. Poller and J. Albert, *Catal. Today*, 2024, 114899.
10. N. Samart, Z. Arhouma, S. Kumar, H. A. Murakami, D. C. Crick and D. C. Crans, *Front. Chem.*, 2018, **6**, 1–16.
11. D. J. Sures, P. I. Molina, P. Miró, L. N. Zakharov and M. Nyman, *New J. Chem.*, 2016, **40**, 928–936.
12. E. E. Nikishina, E. N. Lebedeva and D. V. Drobot, *Inorg. Mater.*, 2012, **48**, 1243–1260.
13. P. Müscher-Polzin, C. Näther and W. Bensch, *Z. Naturforsch.*, 2020, **75**, 583–588.

3 Synthetic Concepts

3.1 Where Are the Elements From?

The most prominent elements in POM chemistry are the group five (V, Nb and Ta) and group six elements (Mo and W) from the periodic table. To understand POM chemistry, it is also important to know where the precursor compounds come from.

3.1.1 Vanadium

Vanadium is found as ores in the lithosphere. Prominent examples are patronite VS_2 (found in Peru), vanadinite $Pb_5[VO_4]_3Cl$ (found in Mexico), roscoelite $K(Al,V)_2(OH,F)_2[AlSi_3O_{10}]$ (found in Colorado), and carnotite $K[UO_2][VO_4]$. In general, vanadium is found in South Africa, China, Russia and the USA. To obtain the metal, in the first step V_2O_5 is precipitated from the ores using H_2SO_4.[1] Eqn (3.1) shows the reaction equation for the generation of V_2O_5 from vanadinite:

$$2Pb_5[VO_4]_3Cl + 10H_2SO_4 \rightarrow 3V_2O_5 + 10PbSO_4 + 2HCl + 9H_2O \tag{3.1}$$

V_2O_5 can be directly reduced to the metal in the second step using calcium (see eqn (3.2)) or converted to the POM precursor $NaVO_3$ in basic pH media (see eqn (3.3)).

$$V_2O_5 + 5Ca \xrightarrow{\sim 950\ ^\circ C} 2V + 5CaO \tag{3.2}$$

$$V_2O_5 + 2NaOH \rightarrow 2NaVO_3 + H_2O \tag{3.3}$$

RSC Foundations No. 3
Polyoxometalate Chemistry
By Jan-Christian Raabe

Published by the Royal Society of Chemistry, www.rsc.org

3.1.2 Niobium and Tantalum

Niobium and tantalum are found as iron or manganese niobate/tantalate (Fe,Mn)$[MO_3]_2$ (with M = Nb or Ta) in Brazil or Canada. In the first step, the ores are digested with HF to obtain the fluorides.[1] Eqn (3.4) shows the reaction equation for the digestion of Nb/Ta ores by HF:

$$(Fe,Mn)[MO_3]_2 + 16HF \rightarrow 4H^+ + 2[MF_7]^{2-} + Fe,MnF_2 + 6H_2O \quad (3.4)$$

The niobium and tantalum fluorides are separated by fractional crystallization of the K^+ salts K_2MF_7 (K_2NbF_7 is easily soluble in water, K_2TaF_7 is poorly soluble). An alternative method to separate both fluorides is a fractional extraction of both components using ketones. With methyl isobutyl ketone, the niobium fluoride can be extracted first and, after lowering the pH value, the tantalum fluoride is extracted into the organic phase. From the fluorides the corresponding element (using elemental Na, see eqn (3.5)) or the oxide can be generated (by the addition of ammonia NH_3 to the aqueous phase containing the fluorides, see eqn (3.6)).[1]

$$[MF_7]^{2-} + 7Na \rightarrow M + 7NaF \quad (3.5)$$

$$4K^+ + 2\,[MF_7]^{2-} + 5H_2O + 10NH_3 \rightarrow M_2O_5 + 4KF + 10NH_4F \quad (3.6)$$

In POM chemistry the niobium/tantalum oxide is used as a precursor (*e.g.* basic melt synthesis).[1]

An alternative approach is the chlorination of the ores to the pentachlorides, which can be separated by fractional distillation. The elements can be obtained by reduction of the chlorides using elemental Na at ~800 °C.[1]

3.1.3 Molybdenum and Tungsten

Both elements are found as oxides in the lithosphere. The most important ore for molybdenum is molybdenite MoS_2 (molybdenum(IV) sulfide) found in Colorado (USA), Norway, Canada, Chile and Germany. In Kärnten (Oberbayern), wulfenite $PbMoO_4$ is found and, rarely, powellite Ca(Mo,W)O_4. For tungsten, the most popular ores are wolframite (Mn,Fe)WO_4 (a mixture of hübnerite $MnWO_4$ and ferberite $FeWO_4$), scheelite $CaWO_4$, stolzite $PbWO_4$ and tuneptite $WO_3{\cdot}H_2O$. The ores are found in China, USA, Canada, South Korea, Bolivia, Portugal, Germany (Erzgebirge) and Russia.[1]

The ores of molybdenum and tungsten are first converted to the oxides MO_3 (with M = Mo and W). For molybdenum, the MoS_2 is directly oxidized to MoO_3 (see eqn (3.7)) and for tungsten the (Mn,Fe) WO_4 is converted with soda to Na_2WO_4 (see eqn (3.8)), which is directly precipitated to the oxide-hydrate $WO_3{\cdot}H_2O$ (see eqn (3.9)).[1]

$$2MoS_2 + 7O_2 \rightarrow 2MoO_3 + 4SO_2 \tag{3.7}$$

$$6(Fe,Mn)WO_4 + 6Na_2CO_3 + O_2 \rightarrow 2(Fe,Mn)_3O_4 + 6Na_2WO_4 + 6CO_2 \tag{3.8}$$

$$Na_2WO_4 + 2HCl \rightarrow WO_3{\cdot}H_2O + 2NaCl \tag{3.9}$$

The metals can be obtained directly from the oxides by reduction with hydrogen (see eqn (3.10)). Alternatively, the oxides can be converted directly to the (mono) metalates (using, for example, sodium hydroxide NaOH), which act as precursor compounds in POM chemistry (see eqn (3.11)).[1]

$$MO_3 + 3H_2 \xrightarrow{800\text{–}1000\ ^\circ C} M + 3H_2O \tag{3.10}$$

$$MO_3 + 2NaOH \rightarrow Na_2MO_4 + H_2O \tag{3.11}$$

3.2 Synthetic Strategies for the Synthesis of Polyoxometalates

In general, there are two strategies known for synthesizing POMs:

1. The self-assembly approach.
2. The lacunary approach.

For the self-assembly approach, all precursor compounds are mixed in their desired stoichiometry (*e.g.* for synthesizing the compound $[PMo_{12}O_{40}]^{3-}$, the stoichiometry of P and Mo precursor compounds needs to be 1 : 12) and are converted to the POM structure by adjusting the pH value and temperature. A typical example is the preparation of the Keggin-type anions $[PM_{12}O_{40}]^{3-}$ with M = Mo, W (see eqn (3.12)–(3.14)):[2–4]

$$12MoO_3 + H_3PO_4 \rightarrow H_3[PMo_{12}O_{40}] \tag{3.12}$$

$$12Na_2MoO_4 + H_3PO_4 + 21HCl \rightarrow Na_3[PMo_{12}O_{40}] + 21NaCl + 12H_2O \tag{3.13}$$

$$12Na_2WO_4 + H_3PO_4 + 21HCl \rightarrow Na_3[PW_{12}O_{40}] + 21NaCl + 12H_2O \tag{3.14}$$

Note: the preparation of $H_3[PW_{12}O_{40}]$ is not possible starting from WO_3. In general, a reaction equation for synthesizing a POM has the general form:

$$\text{precursor1} + (\text{precursor2} + \ldots)\ \text{acid} \rightarrow \text{POM product} + \text{salt} + \text{water}$$

The lacunary approach is based on the following steps:[5–9]

1. Synthesizing the lacunary-type structure *via* self-assembly or starting from an intact parent POM structure (here the intact parent POM structure, *e.g.* a Keggin- or Wells–Dawson-type, is treated in basic pH media to form the lacunary-type structure).
2. The precursor compounds are added to the lacunary-type species.
3. The reaction mixture is acidified to regenerate the fully intact parent POM structure type.

The lacunary approach is a good strategy for synthesizing foreign element-substituted POMs, as discussed later. If the precursors (added in the second step) are precursors of a foreign transition element, the resulting POMs are classified as transition metal-substituted POMs (TMSPOMs). Both strategies are summarized in Figure 3.1.[5–10]

Note:

- TMSPOMs can also be synthesized by the self-assembly approach, as discussed later.
- The precursors that are added to the lacunary-type species in the second step do not necessarily have to be precursors of a foreign transition metal (*e.g.* the anion $[PM_{12}O_{40}]^{3-}$ can be regenerated by adding MO_4^{2-} to the lacunary-type structure $[PM_9O_{34}]^{9-}$).

A big advantage of the self-assembly approach is that the procedure is robust against temperature and pH. This means that temperature and pH fluctuations do not have a negative effect on the result of the procedure. If TMSPOMs are synthesized (as discussed in Chapter 3, Sections 3.5 and 3.6), high degrees of substitution can be achieved.[2–4] A disadvantage of the lacunary approach is that the lacunary-type structure is often only stable in a small pH window. Only low degrees of substitution can be achieved with a foreign element, as the maximum degree of substitution is equal to the number of vacancies of the lacunary-type structure. Both approaches have the disadvantage that the POMs are obtained as alkali salts if alkali metal-containing

self-assembly approach

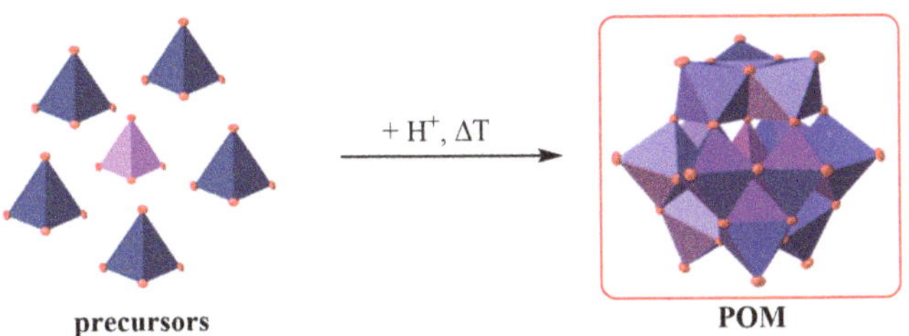

Lacunary approach

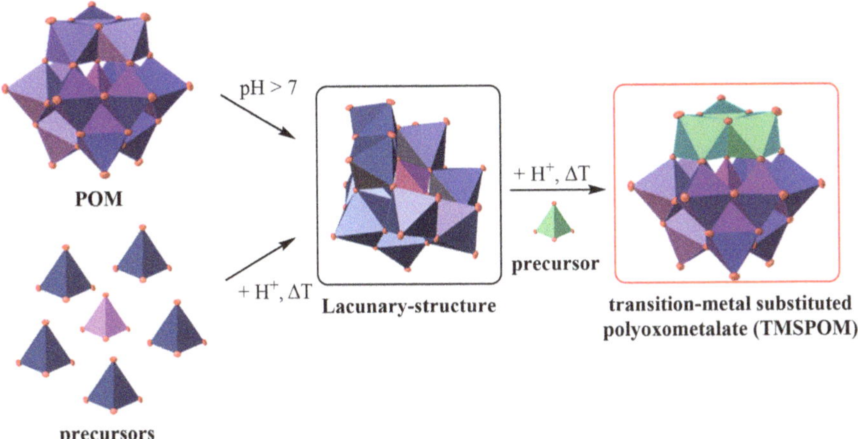

Figure 3.1 The two strategies for synthesizing POMs. In the self-assembly approach, all precursors are added in the correct stoichiometric ratio and the POM structure is formed by adjusting the pH value and/or the temperature.[2-4] In the lacunary approach, a lacunary-type structure is generated from an intact POM structure type in basic pH media or by acidifying precursor compounds (self-assembly approach). In the second step, precursor compounds are added and the vacancies of the lacunary-type structure are filled. The fully intact POM structure type is regenerated by acidification of the reaction system. If the precursors that are added in the second step of the reaction sequence are precursors of a foreign transition element, so-called transition metal-substituted POMs (TMSPOMs) are obtained.[5-10]

precursor compounds are used. In this case, no free POM acids can be obtained, which can be disadvantageous for some applications, as discussed in Chapter 5.[5-10]

Another big advantage of both approaches is that POMs can also be synthesized on a large scale. This is particularly useful if a POM-based application (*e.g.* POM catalysis, see Chapter 5) is to be transferred to the industrial scale.

3.3 Examples of Polyoxometalate Synthesis

Typical equations for the preparation of HPA structures are shown below.

The self-assembly approach can be considered because all the protonated precursors arrange themselves into the final POM structure, driven by thermodynamic driving forces. Important parameters for controlling the synthesis success are the pH value and the temperature. The first parameter plays a very important role, as the POM structure type can rearrange depending on the pH value.[11–13]

Typical precursor compounds for the framework elements are the anions MO_4^{2-} (with M = Mo, W) or VO_3^-, as well as the oxides MoO_3, Nb_2O_5, Ta_2O_5. Phosphoric acid H_3PO_4, metasilicate SiO_3^{2-}, telluric acid $Te(OH)_6$ and potassium hexahydroxoantimonate(v) $K[Sb(OH)_6)]$ are possible precursors for the heteroelements in HPA structures.[14–18] Typical reaction equations for synthesizing HPA structures follow, see eqn (3.15)–(3.18).

3.3.1 Keggin-type Synthesis

$$12Na_2MoO_4 + H_3PO_4 + 21HCl \xrightarrow{\text{pH 1, 20 °C}} Na_3[PMo_{12}O_{40}] + 21NaCl + 12H_2O \tag{3.15}$$

$$12Na_2WO_4 + H_3PO_4 + 21HCl \xrightarrow{\text{pH 1, 20 °C}} Na_3[PW_{12}O_{40}] + 21NaCl + 12H_2O \tag{3.16}$$

$$12MoO_3 + H_3PO_4 \xrightarrow{\text{pH 1, reflux}} H_3[PMo_{12}O_{40}] \tag{3.17}$$

$$Na_3[PM_{12}O_{40}] + 3HCl \xrightarrow{\text{etherate method}} H_3[PM_{12}O_{40}] + 3NaCl \tag{3.18}$$

- Keggin-type POMs are synthesized by combining the precursors for the framework elements (Mo or W) with a precursor for the heteroelement, ideally in a stoichiometric ratio (M : X) of 12 : 1. The precursors for the framework elements are the anions MoO_4^{2-} or WO_4^{2-} (salts are normally purchased as dihydrates).[19] For the heteroelement, the anions PO_4^{3-} or SiO_3^{2-} can be used in order to obtain compounds with stoichiometric ratios of $[PM_{12}O_{40}]^{3-}$ or $[SiM_{12}O_{40}]^{4-}$.[14–18] All combined precursors are acidified to pH 1 (*e.g.* with hydrochloric acid HCl). The heteroelement anion acts as a template around which the metals are arranged to form the structure.[9,20–22] Here, the obtained compounds are called phosphomolybdate/-tungstate (in the case of heteroelement P) or silicomolybdate/-tungstate (in the case of heteroelement Si). For Mo-based Keggin-type phosphomolybdate structures it is possible to use MoO_3 as framework element precursor, which can be dissolved in diluted H_3PO_4 solution by refluxing.[2–4] A big advantage is that the POM is isolated as POM acid $H_3[PMo_{12}O_{40}]$ and not as alkali salt.[23,24] The advantage of using POM acids in catalytic applications is discussed later. Here, the compound is called molybdatophosphoric acid or phosphomolybdic acid.[2–4,25]

As described above, if alkali precursors like MO_4^{2-} are used, all POMs are isolated as alkali salts, *e.g.* $Na_3[PM_{12}O_{40}]$, and alkali halogenides are obtained as by-products.[23,24] Nevertheless, it is possible to produce POM acids from the POM alkali salts. One possibility is the so-called etherate method.[26,27] This method is often referred to as extraction, but only a complexation of the POM takes place. In practice, diethyl ether is added to the aqueous, acidified POM phase, which contains the alkali cations and the POM anion. After mixing the two phases, a third phase with a density $>1\ g\,cm^{-3}$ is formed, which settles below the aqueous phase. In this phase, the POM anion is complexed by the ether and is protonated by excess protons in the aqueous phase. All alkali cations remain in the aqueous phase as alkali halides. The POM acid can then be obtained from the POM ether phase by evaporating the ether. However, the success of the etherate method is limited to low-charged POM anions, as the POM-ether complex forms only with POM anions of low anionic charges.[26–28]

3.3.2 Wells–Dawson-type Synthesis

$$18Na_2WO_4 + 2H_3PO_4 + 30HCl \xrightarrow{\text{pH}<2,\ \text{reflux}} Na_6[\alpha\text{-}P_2W_{18}O_{62}] + 30NaCl + 18H_2O \quad (3.19)$$

$$Na_6[\alpha\text{-}P_2W_{18}O_{62}] + 6KCl \xrightarrow{20\ ^\circ\text{C}} K_6[\alpha\text{-}P_2W_{18}O_{62}] + 6NaCl \quad (3.20)$$

$$18Na_2WO_4 + 2H_3PO_4 + 30HCl \xrightarrow{\text{pH}<2,\ \text{reflux, 4 hours}} Na_6[\alpha/\beta\text{-}P_2W_{18}O_{62}] + 30NaCl + 18H_2O \quad (3.21)$$

$$Na_6[\alpha/\beta\text{-}P_2W_{18}O_{62}] + 6NH_4Cl \xrightarrow{20\ ^\circ\text{C}} (NH_4)_6[\alpha/\beta\text{-}P_2W_{18}O_{62}] + 6NaCl \quad (3.22)$$

$$(NH_4)_6[\alpha/\beta\text{-}P_2W_{18}O_{62}] \xrightarrow{45\ ^\circ C,\ 5\ days} (NH_4)_6[\beta\text{-}P_2W_{18}O_{62}] \quad (3.23)$$

$$(NH_4)_6[\alpha\text{-}P_2W_{18}O_{62}] + 6KCl \xrightarrow{20\ ^\circ C,\ NH_4Cl} K_6[\alpha\text{-}P_2W_{18}O_{62}] + 6(NH_4)Cl \quad (3.24)$$

- Wells–Dawson-type structures can be synthesized according to Israel-Martyr Mbomekalle[29] by acidification of WO_4^{2-} solutions and by adding H_3PO_4. The solution is refluxed for 24 hours (see eqn (3.19)).[29] The ideal stoichiometric ratio (M : X) is 18 : 2 or 9 : 1. A cation-exchange is performed by adding potassium chloride KCl (see eqn (3.20)). This has the advantage that the K^+ salt of the Wells–Dawson-type structure is easier to isolate by crystallization than the pure Na^+ salt.[29] A second approach was published by Roland Contant (*Inorganic Syntheses*, 1990, Volume 27).[30] Here, both isomers of the anion $[P_2W_{18}O_{62}]^{6-}$ (α and β) are formed, as shown in eqn (3.21) (isomers of POMs are discussed in Chapter 4, Section 4.3), which are precipitated as NH_4^+ salt as an isomer mixture (see eqn (3.22)). Elemental bromine can be used to reoxidize W(v) to W(vi) if any reduced species are available. Only the β isomer can be crystallized if the isomer mixture is dissolved in warm water with subsequent slow evaporation of water after five days. The pure α isomer can be precipitated from the remaining, aqueous solution as K^+ salt, as shown in eqn (3.24).[30] It should be noted the procedure of Mbomekalle only produces the α isomer.[29] However, the synthesis of a Wells–Dawson-type structure is much more difficult than that of a Keggin-type structure. Scientists have been working for many years to find the ideal synthesis conditions.[31] Mo-based Wells–Dawson structures of the type $[P_2Mo_{18}O_{62}]^{6-}$ are also known. However, Mo-based Wells–Dawson-type structures prove to be less stable and are therefore less explored.[29,30,32–35]

3.3.3 Anderson–Evans-type Synthesis

$$6Na_2MoO_4 + Te(OH)_6 + 6HCl \xrightarrow{pH\ 5,\ 20\ ^\circ C} Na_6[TeMo_6O_{24}] + 6NaCl + 6H_2O$$
$$6Na_2WO_4 + Te(OH)_6 + 6HCl \xrightarrow{pH\ 5,\ 20\ ^\circ C} Na_6[TeW_6O_{24}] + 6NaCl + 6H_2O \quad (3.25)$$

- Compared to the Wells–Dawson-type structure, the experimental procedure for synthesizing an Anderson–Evans-type structure requires only a MO_4^{2-} precursor, which is combined with a heteroelement precursor, *e.g.* $Te(OH)_6$ (X = Te) or $K[Sb(OH)_6]$ (X = Sb), as shown in eqn (3.25). The aqueous solution is only acidified to pH 5 in order to prevent precipitation of WO_3 or MoO_3.[36–38]

3.3.4 Lindqvist-type Synthesis

Typical equations for synthesizing Lindqvist-type IPA structures are shown below.

$$6Na_2MoO_4 + 10HCl \xrightarrow{\text{pH 1, 20 °C}} Na_2[Mo_6O_{19}] + 10NaCl + 5H_2O$$

$$6Na_2WO_4 + 10HCl \xrightarrow{\text{pH 1, 80 °C}} Na_2[W_6O_{19}] + 10NaCl + 5H_2O \quad (3.26)$$

$$3Nb_2O_5 + 8KOH \xrightarrow{300\ °C} K_8[Nb_6O_{19}] + H_2O$$

$$3Ta_2O_5 + 8KOH \xrightarrow{>300\ °C} K_8[Ta_6O_{19}] + H_2O \quad (3.27)$$

- To synthesize Lindqvist-type structures, the precursors MO_4^{2-} are continuously acidified at elevated temperatures (see eqn (3.26)). Different pH values are optimal for different Lindqvist-type POMs. As previously mentioned, the preparation of anions $[Nb_6O_{19}]^{8-}$ and $[Ta_6O_{19}]^{8-}$ requires special conditions, as their synthesis is only possible by basic melt approaches at temperatures above 300 °C from the metal oxides (see eqn (3.27)). Here, the metal oxides, in particular the M–O–M bonds, are cleaved in a basic phase. Since anions of the types MO_4^{3-} are not stable for M = Nb, Ta, the decomposition ends at the level of the hexametalates.[30,39–47]

3.3.5 Dexter–Silverton-type Synthesis

$$(NH_4)_2[Ce(NO_3)_6] + 2(NH_4)_6[Mo_7O_{24}] + 4NH_4NO_3 + 4H_2SO_4 + 2H_2O$$
$$\xrightarrow{100\ °C} (NH_4)_6H_2[CeMo_{12}O_{42}] + 10HNO_3 + 4(NH_4)_2SO_4 + 2(NH_4)_2MoO_4 \quad (3.28)$$

- The synthesis of a Dexter–Silverton-type POM is possible using a self-assembly approach by boiling an aqueous ammonium cerium(IV) nitrate $(NH_4)_2[Ce(NO_3)_6]$ and paramolybdate solution (see eqn (3.28)). Presumably, the H_2SO_4 serves to initiate the self-assembly mechanism and acts as an additional proton source to obtain mixed salts, like $(NH_4)_6H_2POM$. Ammonium nitrate NH_4NO_3 could act as an additional source of NH_4^+ ions, since two MoO_4^{2-} ions have to be eliminated stoichiometrically.[48]

3.3.6 Waugh-type Synthesis

$$(NH_4)_6[Mo_7O_{24}] \xrightarrow[95\ ^\circ C]{Mn^{II}SO_4,\ (NH_4)_2S_2O_8} (NH_4)_6[Mn^{IV}Mo_9O_{32}]$$

$$via\text{: } Mn^{II}SO_4 + (NH_4)_2S_2O_8 \xrightarrow{in\ situ} Mn^{IV}(SO_4)_2 + (NH_4)_2SO_4 \quad (3.29)$$

$$10Na_2MoO_4 \xrightarrow{1.\ 20\ ^\circ C,\ H_3PO_4,\ V_2O_5;\ 2.\ 100\ ^\circ C;\ 3.\ 20\ ^\circ C} Na_4[Mo_{10}O_{32}] \quad (3.30)$$

- Waugh-type synthesis is possible using a two-step approach, following a self-assembly route. First, manganese(II) sulfate $MnSO_4$ is oxidized with ammonium peroxodisulphate $(NH_4)_2S_2O_8$ in aqueous solution. The oxidized Mn(IV) species is added to an aqueous ammonium paramolybdate $(NH_4)_6[Mo_7O_{24}]$ solution at 95 °C in order to obtain $[MnMo_9O_{32}]^{4-}$ (see eqn (3.29)).[49,50] An experimental procedure to prepare an IPA Waugh-type structure, a compound which is also known as decamolybdate $[Mo_{10}O_{32}]^{4-}$, was reported by C. W. Hu. In this procedure, an aqueous MoO_4^{2-} solution is acidified with H_3PO_4 and V_2O_5 is added (see eqn (3.30)). The role of V_2O_5 is unknown, but without V_2O_5 the procedure does not work. PO_4^{3-} and V(V) species are not found in the target product.[51]

3.3.7 Strandberg-type Synthesis

$$5Na_2MoO_4 + 2NaH_2PO_4 + 6HClO_4 \xrightarrow{20\ ^\circ C,\ days/weeks} Na_6[P_2Mo_5O_{23}] + 6NaClO_4 + 5H_2O \quad (3.31)$$

- Strandberg-type POMs are obtained following a self-assembly mechanism by acidifying an aqueous Na_2MoO_4 and dihydrogen phosphate NaH_2PO_4 solution with perchloric acid $HClO_4$ at room temperature (see eqn (3.31)). The crystals are obtained after some days or weeks.[52]

3.3.8 $[P_4W_{14}O_{58}]^{12-}$-type Synthesis

$$14Na_2WO_4 + 4H_3PO_4 + 16HCl \xrightarrow{20\ ^\circ C} Na_{12}[P_4W_{14}O_{58}] + 14H_2O + 16NaCl \tag{3.32}$$

- The anion $[P_4W_{14}O_{58}]^{12-}$ is synthesized according to a procedure published in 2024,[38] which follows a self-assembly approach. Here, an aqueous WO_4^{2-} and PO_4^{3-} solution is acidified to pH 5. The compound is isolated as Na^+ salt (see eqn (3.32)).[38]

3.3.9 Preyssler-type Synthesis

$$30Na_2WO_4 + 20H_3PO_4 \xrightarrow[]{\substack{>1\ \text{bar} \\ 120\ ^\circ C}} Na_{15}[P_5W_{30}O_{110}] + 30H_2O + 15Na_3PO_4 \tag{3.33}$$

$$30Na_2WO_4 + 20H_3PO_4 \xrightarrow[]{\substack{100\ ^\circ C \\ 5\ \text{hours}}} Na_{15}[P_5W_{30}O_{110}] + 30H_2O + 15Na_3PO_4 \tag{3.34}$$

- The Preyssler-type anion can be prepared under hydrothermal conditions at 120 °C, in which an aqueous WO_4^{2-} solution is treated with H_3PO_4, or by refluxing an aqueous WO_4^{2-} and H_3PO_4 solution for five hours. Both approaches follow a self-assembly mechanism.[30,53,54] The Preyssler-type anion is also formed as a by-product in the Wells–Dawson-type synthesis, according to Contant (see eqn (3.33) and (3.34)).[30]

3.3.10 14-Heteropolyvanadate Synthesis

$$14NaVO_3 + 3H_3PO_4 \xrightarrow{20\ ^\circ C} Na_9[PV_{14}O_{42}] + Na_3PO_4 + Na_2HPO_4 + 4H_2O \tag{3.35}$$

$$V_2O_5 \xrightarrow{0\ ^\circ C,\ H_2O_2} VO(O_2)^+ + \text{peroxy anions of V(v)} \xrightarrow[-O_2]{} \underset{\text{Decavanadate}}{H_6[V_{10}O_{28}]} \tag{3.36}$$

$$H_6[V_{10}O_{28}] \xrightarrow{0\ ^\circ C,\ H_2O_2} H_9[PV_{14}O_{42}] \tag{3.37}$$

- The 14-heteropolyvanadate anion can be prepared, according to Kato, Kobayashi and Sasaki, by acidifying an aqueous VO_3^- solution with H_3PO_4. Here, species $[PV_{14}O_{42}]^{9-}$ is obtained as Na^+ salt (see eqn (3.35)). A cation-exchange with organic cations can be done by adding organic ammonium compounds like amines or guanidine.[55] A second approach was introduced by V. F. Odyakov.[2,3] Here, the experimental procedure uses V_2O_5, which can be dissolved in aqueous, diluted H_2O_2 solution. Peroxo compounds of V(V) are formed, which release molecular O_2 to form the well-known decavanadate anion $[V_{10}O_{28}]^{6-}$ (see above). The compound is obtained as POM acid $H_6[V_{10}O_{28}]$, which can be converted directly to the 14-heteropolyvanadate $H_9[PV_{14}O_{42}]$ species by adding H_3PO_4 (see eqn (3.36) and (3.37)).[2,3]

Further IPA synthesis approaches are shown below.

3.3.11 Decatungstate Synthesis

$$10Na_2WO_4 + 16HCl \xrightarrow{20\ ^\circ C,\ pH\ 2} Na_4[W_{10}O_{32}] + 16NaCl + 8H_2O \quad (3.38)$$

- Decatungstate is synthesized using a self-assembly approach starting from an aqueous solution of WO_4^{2-}, which is acidified to pH values of 2 in order to prevent WO_3 precipitation at more acidic pH values of <1.5 (see eqn (3.38)).[56–58]

3.3.12 Weakley-Yamase-type Synthesis

$$10Na_2WO_4 + Eu(NO_3)_3 + 8CH_3COOH \xrightarrow[pH\ 6.5\text{–}7.5]{60\text{–}100\ ^\circ C} Na_9[Eu(W_5O_{18})_2] + 8CH_3COONa + 3NaNO_3 \quad (3.39)$$

- A Weakley–Yamase-type structure is synthesized using a self-assembly approach at temperatures between 60–100 °C by acidifying an aqueous WO_4^{2-} solution to pH values of between 6.5 and 7.5 using acetic acid (AA). To this solution, an aqueous lanthanide nitrate or chloride solution is added. The salt can be crystallized from the solution at low temperatures (~5 °C), see eqn (3.39).[59,60]

3.3.13 Decavanadate Synthesis

$$10NaVO_3 + 4HCl \xrightarrow{20\ ^\circ C} Na_6[V_{10}O_{28}] + 4NaCl + 2H_2O \quad (3.40)$$

$$V_2O_5 \xrightarrow{0\ ^\circ C,\ H_2O_2} VO(O_2)^+ + \text{peroxy anions of V(V)} \xrightarrow[-O_2]{} H_6[V_{10}O_{28}] \quad (3.41)$$

- Decavanadate is prepared by acidifying aqueous solutions of $NaVO_3$ (see eqn (3.40)). An alternative approach was presented by Odyakov,[2,3] starting from V_2O_5, which can be dissolved in aqueous H_2O_2 solution. Molecular O_2 is released, resulting in the POM acid species $H_6[V_{10}O_{28}]$ (see eqn (3.41)).[2,3]

3.3.14 Paradodecatungstate/Paratungstate B Synthesis

$$12Na_2WO_4 + 16HCl \xrightarrow{20\ ^\circ C,\ pH\ 5} Na_8[W_{12}O_{40}] + 16NaCl + 8H_2O \quad (3.42)$$

- Paradodecatungstate/paratungstate B can be synthesized by acidifying an aqueous WO_4^{2-} solution to pH 5, following a self-assembly mechanism (see eqn (3.42)).[38]

3.3.15 Paratungstate (A) Synthesis

- A paratungstate A anion can only be isolated using big cations. Cations and anions must be prepared in two different steps and the final compound (*e.g.* $[Mg_{12}Al_6(OH)_{36}][W_7O_{24}]$) is obtained by an ion-exchange step after aqueous solutions of cation and anion are combined. The anion is prepared according to paratungstate B, using the equilibrium shift of the conversion of paratungstate B into A. In the final step, an ion-exchange is done in order to obtain the salt $[Mg_{12}Al_6(OH)_{36}][W_7O_{24}]$ from the aqueous $Na_7[W_7O_{24}]$ and $[Mg_{12}Al_6(OH)_{36}](NO_3)_6$ solution.[61]

Cation preparation:

$$12Mg(NO_3)_2 + 6Al(NO_3)_3 + 36NaOH \xrightarrow{20\ ^\circ C} [Mg_{12}Al_6(OH)_{36}](NO_3)_6 + 36NaNO_3 \quad (3.43)$$

Anion preparation:

$$7Na_2WO_4 + 8HNO_3 \xrightarrow{20\ ^\circ C,\ pH\ 6\text{–}7} Na_6[W_7O_{24}] + 8NaNO_3 + 4H_2O \quad (3.44)$$

Combining eqn (3.43) and (3.44) results in the final reaction equation eqn (3.45):

$$Na_6[W_7O_{24}] + [Mg_{12}Al_6(OH)_{36}](NO_3)_6 \xrightarrow{\substack{1.\ 20\ ^\circ C \\ 2.\ 73\text{–}74\ ^\circ C \\ 36\ h \\ 3.\ aging,\ 24\ h}} [Mg_{12}Al_6(OH)_{36}][W_7O_{24}] + 6NaNO_3 \quad (3.45)$$

3.3.16 Further IPA Anion Synthesis

The syntheses of the following IPA anions should also be discussed.

- The paramolybdate anion, normally known as hexaammonium salt, is obtained according to a self-assembly route by dissolving

MoO_3 in aqueous NH_3 solution (eqn (3.46)). It is also an intermediate by acidifying aqueous MoO_4^{2-} solutions (pH > 4).

$$7MoO_3 + 6NH_3 + 3H_2O \xrightarrow{20\ °C} (NH_4)_6[Mo_7O_{24}] \quad (3.46)$$

- Octamolybdate is synthesized using a self-assembly procedure by acidifying an aqueous MoO_4^{2-} solution to pH 4, see eqn (3.47).[62]

$$8Na_2MoO_4 + 12HCl \xrightarrow{20\ °C,\ pH\ 4} Na_4[Mo_8O_{26}] + 12NaCl + 6H_2O \quad (3.47)$$

- The anion $[Mo_{36}O_{112}(H_2O)_{16}]^{8-}$ can be synthesized using a self-assembly approach by acidifying an aqueous MoO_4^{2-} solution with nitric acid HNO_3. In the original procedure of Bernt Krebs, the compound was obtained as K^+ salt, using K_2MoO_4 as precursor (see eqn (3.48)).[63]

$$36K_2MoO_4 + 64HNO_3 \xrightarrow{20\ °C} K_8[Mo_{36}O_{112}(H_2O)_{16}] + 64KNO_3 + 16H_2O \quad (3.48)$$

- The $\{Mo_{154}\}$ rings are obtained by acidifying an aqueous MoO_4^{2-} solution in the presence of a reducing agent, like thiosulfate $Na_2S_2O_3$, following a self-assembly mechanism (eqn (3.49)). After one day, the crystals of the Na^+ salt can be collected by filtration.[64]

$$Na_2MoO_4 \xrightarrow[1\ day]{Na_2S_2O_4,\ HCl} Na_{15}\left[Mo^{VI}_{126}Mo^{V}_{28}O_{462}H_{14}(H_2O)_{70}\right] \quad (3.49)$$

The examples shown here illustrate that most of the POM syntheses are based on the self-assembly approach. Examples in which lacunary approaches are used follow in the next subchapters. All approaches presented here are only examples and should give the reader a feeling for preparing different types of POM structures.

3.4 Examples of Lacunary-type Polyoxometalate Synthesis

Lacunary-type POMs can be prepared by two routes:

1. In a self-assembly approach.
2. Starting from an intact POM structure type, which is treated under basic conditions. Some of the MO_6 octahedra are removed, forming the vacancies.

The most important lacunary-type structures are the compounds that are formed from the Keggin- and Wells–Dawson-type structures. For Keggin-type POMs, a mono $[XM_{11}O_{39}]^{7-}$ and a three-fold lacunary-type $[XM_9O_{34}]^{9-}$ species are known. In special cases there is also a two-fold lacunary-type species, like $[XM_{10}O_{36}]^{8-}$. For the Wells–Dawson-type structure there is a mono $[X_2M_{17}O_{61}]^{10-}$, three- $[X_2M_{15}O_{56}]^{12-}$ and six-fold lacunary-type anion $[H_2P_2W_{12}O_{48}]^{12-}$.

3.4.1 Keggin Lacunary Synthesis

Typical reaction equations for synthesizing lacunary-type POMs are shown in eqn (3.50) and (3.52)–(3.55). Keggin lacunary-type anions with three vacancies $[XMo_9O_{34}]^{9-}$ can be synthesized in a self-assembly approach according to Sadayuki Himeno, starting from $MoO_4{}^{2-}$ and $PO_4{}^{3-}$ (see eqn (3.50)).[9,10,32] For the anion $[PMo_9O_{34}]^{9-}$, a side reaction is known in which a dimerization to form the Wells–Dawson-type structure $[P_2Mo_{18}O_{62}]^{6-}$ occurs if an aqueous solution is heated under pH acidic conditions ($\Delta G = -357\ \text{kJ mol}^{-1}$), see eqn (3.51).[10,65] Synthetic approaches for the anion $[PW_9O_{34}]^{9-}$ are based on the work of Domaille and Watunya. In contrast to the anion $[PMo_9O_{34}]^{9-}$, $[PW_9O_{34}]^{9-}$ precipitates directly from the aqueous solution and can be isolated by filtration.[66] Approaches for preparing Keggin lacunary-type structures with only one vacancy are possible according to both methods.[18,19,66–70]

$$9Na_2MoO_4 + Na_2HPO_4 + 11HCl \xrightarrow{\text{pH 1, 20 °C}} Na_9[PMo_9O_{34}] + 11NaCl + 6H_2O \quad (3.50)$$

Side reaction:

$$2Na_9[PMo_9O_{34}] + 12HCl \xrightarrow{\text{pH 1, 40–100 °C}} Na_6[P_2Mo_{18}O_{62}] + 12NaCl + 6H_2O \quad (3.51)$$

$$9Na_2WO_4 + H_3PO_4 + 9HOAc \xrightarrow{\text{pH 7–8, 20 °C}} Na_9[PW_9O_{34}] + 9NaOAc + 6H_2O \quad (3.52)$$

$$H_3[PMo_{12}O_{40}] + 9NaOH \xrightarrow{\text{pH 4.3, 20 °C}} Na_7[PMo_{11}O_{39}] + Na_2MoO_4 + 6H_2O^{*} \quad (3.53)$$

$$H_3[PW_{12}O_{40}] + 9KOH \xrightarrow{\text{pH 5}} K_7[PW_{11}O_{39}] + K_2WO_4 + 6H_2O^{**} \quad (3.54)$$

$$11Na_2WO_4 + Na_2HPO_4 + 17HNO \xrightarrow{\text{pH 4–5, 80 °C}} Na_7[PW_{11}O_{39}] + 17NaNO_3 + 9H_2O \quad (3.55)$$

* NaOH *via* $NaHCO_3$; ** KOH *via* $KHCO_3$

3.4.2 Wells–Dawson Lacunary Synthesis

$$Na_6[P_2W_{18}O_{62}] + 6NaOH \xrightarrow{\text{pH}>7,\ 20\ °C} Na_{10}[P_2W_{17}O_{61}] + Na_2WO_4 + 3H_2O^{*} \quad (3.56)$$

$$Na_6[P_2W_{18}O_{62}] + 12NaOH \xrightarrow{\text{pH 9, 20 °C}} Na_{12}[P_2W_{15}O_{56}] + 3Na_2WO_4 + 6H_2O^{***} \quad (3.57)$$

$$(NH_4)_6[P_2W_{18}O_{62}] + 12(NH_4)OH + 6NH_4Cl + 6(HOCH_2)_3CNH_2 \xrightarrow{\text{pH 8.3–8.6, 20 °C}} (NH_4)_{12}[H_2P_2W_{12}O_{48}] + 6(NH_4)_2WO_4 + 6(HOCH_2)_3CNH_3^{+} + 6Cl^{-} + 2H_2O \quad {****} \quad (3.58)$$

* NaOH *via* $NaHCO_3$; *** NaOH *via* Na_2CO_3; **** $(NH_4)OH$ *via* $(NH_4)_2CO_3$

Wells–Dawson lacunary-type structures can only be prepared starting from the original Wells–Dawson-type anion $[P_2W_{18}O_{62}]^{6-}$, which is treated under basic pH conditions. It should be emphasized that the selectivity of lacunary-type formation can be shifted to the anion $[P_2W_{17}O_{61}]^{10-}$ or $[P_2W_{15}O_{56}]^{12-}$ by selecting the bases sodium hydrogen carbonate $NaHCO_3$ or sodium carbonate Na_2CO_3 and is depending on the base strength (see eqn (3.56) and (3.57)). After basification the final lacunary-type species crystallizes from the aqueous solution.[30,32,67,71–74]

The six-fold Wells–Dawson lacunary-type anion is prepared from the NH_4^+ salt of the classical Wells–Dawson-type anion, which is treated with ammonium carbonate $(NH_4)_2CO_3$. Ammonium chloride NH_4Cl acts as a source of additional NH_4^+ cations and the base tris(hydroxymethyl)aminomethane captures the additional protons that are produced to ensure that the pH value does not decrease (see eqn (3.58)).

With the concepts for preparing lacunary-type structures, a powerful concept is now available for synthesizing TMSPOMs, which will be discussed in the next subchapter.

3.5 Transition Metal-substituted Polyoxometalates

Transition metal substitution means substituting one or more framework elements in a POM structure with foreign transition metals. This is a current field of research, as transition metal substitution gives the POM new properties, including changes in

- the electronic structure of the POM (*e.g.* reduction/oxidation, RedOx activity);
- the overall charge of the cluster (if the foreign element has a different oxidation state);
- the spectroscopic and crystallographic data; and
- the acidity, if the final cluster is a POM acid.

The first and last point are very important for catalytic applications, as discussed in Chapter 5.

The best-known examples of TMSPOMs are the V(v) substituted Keggin-, Wells–Dawson- and Lindqvist-type structures, with their molecular and stoichiometric formulas $[PV_xMo_{12-x}O_{40}]^{(3+x)-}$ ($x=1$ to 6), $[P_2V_xW_{18-x}O_{62}]^{(6+x)-}$ and $[V_xW_{6-x}O_{19}]^{(2+x)-}$ ($x=1$ to 3). TMSPOMs are synthesized using the self-assembly approach or from lacunary-type structures.

The temporal development of the approaches for preparing V(v) substituted Keggin-type phosphomolybdates is shown in eqn (3.59)–(3.64).

1. Tsigdinos

$$(12-x)Na_2MoO_4 + xNaVO_3 + Na_2HPO_4 + (24.5-1.5x)H_2SO_4 \xrightarrow[\text{pH 1, Reflux}]{\text{etherate method}} H_{3+x}[PV_xMo_{12-x}O_{40}] + (12-x)H_2O + (13-0.5x)Na_2SO_4 \tag{3.59}$$

2. Grate

$$(12-x)Na_2MoO_4 + xNaVO_3 + Na_2HPO_4 + \text{excess } H_2SO_4 \xrightarrow{\text{pH 1, 20 °C}} Na_yH_{(3+x-y)}[PV_xMo_{12-x}O_{40}] + (12-x)H_2O + \text{excess } Na_2H_{2-z}SO_4 \tag{3.60}$$

3. Odyakov

$$H_3PO_4 + (12-x)MoO_3 + \frac{1}{2}xV_2O_5 + \frac{1}{2}xH_2O \xrightarrow[\text{solvent: } H_2O]{\text{1. pH<1, 0 °C; 2. pH<1, reflux}} H_{(3+x)}\left[PV_xMo_{(12-x)}O_{40}\right] \tag{3.61}$$

$$V_2O_5 \xrightarrow[]{\text{1. cooled } H_2O_2 \text{ 2. } H_3PO_4} H_9[PV_{14}O_{42}] \quad (3.62)$$

$$H_3PO_4 + (12 - x)MoO_3 + 0.1xH_6[V_{10}O_{28}] + 0.2xH_2O \xrightarrow[\text{solvent: } H_2O]{} H_{(3+x)}[PV_xMo_{(12-x)}O_{40}] \quad (3.63)$$

$$(12 - x)MoO_3 + \left(1 - \frac{x}{14}\right)H_3PO_4 + \frac{x}{14}H_9[PV_{14}O_{42}] + \frac{2x}{7}H_2O \xrightarrow[\text{solvent: } H_2O]{} H_{(3+x)}[PV_xMo_{(12-x)}O_{40}] \quad (3.64)$$

The chronological development of V(v) substitution in Keggin-type structures goes back to George A. Tsigdinos. He isolated the phosphomolybdate structures as POM acids by combining the salts of the anions MoO_4^{2-}, VO_3^- and PO_4^{3-} in the desired stoichiometry and converting them to the TMSPOM structure by specifically adjusting the pH value. Using the etherate method, the TMSPOMs are isolated as acids. However, it is not possible to prepare POM acids with substitution degrees $x > 3$ with this approach because highly negatively charged POM anions result in the failure of the etherate method.[75] In 1996, John H. Grate was able to isolate the TMSPOMs as mixed acid and alkali salt by reacting the precursors with an excess of H_2SO_4.[76] Finally, the process for the preparation of V(v) substituted Keggin-type TMSPOMs was optimized by Odyakov. It is now possible to isolate POM acids with degrees of substitution of up to $x = 6$. Their strategy avoids using precursors containing alkali metals and only requires metal oxides, such as MoO_3 and V_2O_2.[2–4,77,78]

Here, in the first step, MoO_3 is dissolved in diluted, aqueous H_3PO_4 solution. In the second step, the V(v) precursor is prepared according to eqn (3.62) and in the final step, eqn (3.64), the $H_9[PV_{14}O_{42}]$ precursor is added to the initially formed $H_3[PMo_{12}O_{40}]$ solution to form the final Keggin-type TMSPOM. The target substitution degree is ensured by combining the P : V : Mo precursors in their correct stoichiometric ratios. Here, x cannot be more than 6. For $x = 6$ the anionic charge is -9 and for $x = 7$ it would be -10. Obviously the anionic charge of -10 is too high for those species.[2–4,77,78]

It is also possible to prepare the anions $[PV_xW_{12-x}O_{40}]^{(3+x)-}$. All the substituted TMSPOMs can be prepared *via* the lacunary (up to $x = 3$) or *via* the self-assembly approach ($x = 1$ to 6). Since WO_3 cannot be used as a precursor, these anions can only be prepared as alkali salts, using Na_2WO_4 as a water-soluble alternative. POM acid formation is only possible up to $x = 3$, as the etherate method is limited for higher degrees of substitution.[28]

3.5.1 Lacunary Approach for TMSPOMs

Specific lacunary approaches for preparing Keggin-type TMSPOMs are shown in eqn (3.65)–(3.67).[10,28]

$$Na_9[PMo_9O_{34}] + 3NaVO_3 + 6HCl \xrightarrow{20\ ^\circ C/\text{reflux, pH 1}} Na_6[PV_3Mo_9O_{40}] + 6NaCl + 3H_2O \quad (3.65)$$

$$Na_9[PW_9O_{34}] + 3NaVO_3 + 6HCl \xrightarrow{20\ ^\circ C/\text{reflux, pH 1}} Na_6[PV_3W_9O_{40}] + 6NaCl + 3H_2O \quad (3.66)$$

$$Na_9[PM_9O_{34}] + 2NaVO_3 + Na_2MO_4 + 8HCl \xrightarrow{20\ ^\circ C/\text{reflux, pH 1}} Na_5[PV_2M_{10}O_{40}] + 8NaCl + 4H_2O \quad (3.67)$$

The substitution degree is limited to the number of vacancies of the lacunary-type species (here for $[PM_9O_{34}]^{9-}$ $x \leq 3$).

As introduced in Chapter 1, a Keggin-type cluster has 14 so-called "pits", which can be categorized in three groups—A, B and C—as shown in Figure 3.2.[55]

The pits are reactive and can coordinate additional metal ions. According to the work of Chifeng Li, it is possible to show that the anion $[PV_3Mo_9O_{40}]^{6-}$ can coordinate one more VO^{2+} cation, resulting in a cluster with the stoichiometric ratio of $[PV_3Mo_9O_{40}(VO)]^{3-}$.[6] So, the structure $[PV_{14}O_{42}]^{9-}$ can be interpreted as a Keggin-type species $[PV_{12}O_{40}]^{15-}$ on which two additional VO^{2+} cations are coordinated.[55]

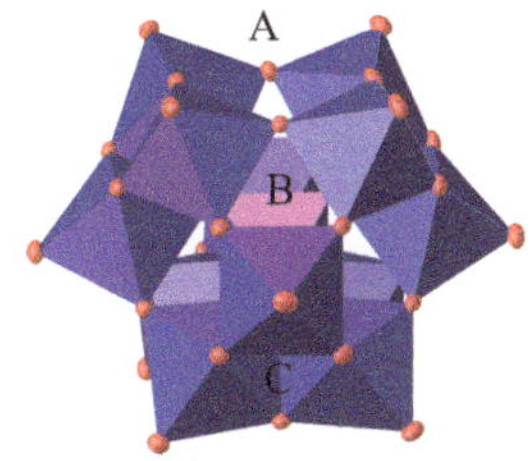

Figure 3.2 So-called "pits" on a Keggin-type structure, which can be classified into group A, B and C. The datawere used from the Cambridge Crystallographic Data Centre and Fachinformationszentrum Karlsruhe AccessStructures service database (deposition number: 2177881).

3.5.2 Wells–Dawson-type TMSPOM Synthesis

V(V) substituted Wells–Dawson-type clusters can only be prepared starting from the lacunary-type structures, which are added to an acidified $NaVO_3$ solution (see eqn (3.68) and (3.69)). The final TMSPOMs are isolated by crystallization due to the addition of KCl.[29,32,71,72,79–84]

$$Na_{10}[P_2W_{17}O_{61}] + NaVO_3 + 4HCl \xrightarrow{pH<2,\ reflux} Na_7[P_2VW_{17}O_{62}] + 4NaCl + 2H_2O \quad (3.68)$$

$$Na_{12}[P_2W_{15}O_{56}] + 3NaVO_3 + 6HCl \xrightarrow{pH<2,\ reflux} Na_9[P_2V_3W_{15}O_{62}] + 6NaCl + 3H_2O \quad (3.69)$$

3.5.3 Lindqvist-type TMSPOM Synthesis

$$(6-x)Na_2WO_4 + xNaVO_3 + (10-2x)HCl \xrightarrow{*} Na_{2+x}[V_xW_{6-x}O_{19}] + (10-2x)NaCl + (5-x)H_2O \quad (3.70)$$

$$4Na_2MoO_4 + 2NaVO_3 + 6HCl \xrightarrow{*} Na_{2+x}[V_xW_{6-x}O_{19}] + (10-2x)NaCl + (5-x)H_2O \quad (3.71)$$

* different pH values, 80 °C.

TMSPOMs based on the Lindqvist-type structure are synthesized according to a self-assembly approach, as shown in the equations above (eqn (3.70) and (3.71)).[39,43] Elevated temperatures and acidic pH conditions of the aqueous precursor solutions are required.

3.6 More Synthetic Strategies to Prepare Polyoxometalates

The synthesis of POMs that will be discussed in Chapter 6 is summarized in this section. These POMs are used in different catalytic or biomedicine applications.

Silicomolybdate/-tungstate Synthesis

$$12Na_2MoO_4 + Na_2SiO_3 + 22HCl \xrightarrow{20\ °C,\ 15\ min} Na_4[SiMo_{12}O_{40}] + 22NaCl + 11H_2O \quad (3.72)$$

$$12Na_2WO_4 + Na_2SiO_3 + 22HCl \xrightarrow{20\,°C} Na_4[SiW_{12}O_{40}] + 22NaCl + 11H_2O \quad (3.73)$$

$$H_4[SiM_{12}O_{40}] \xleftarrow[-4NaCl]{+4HCl \text{ etherate method}} Na_4[SiM_{12}O_{40}] \xrightarrow[-4NaCl]{+4KCl} K_4[SiM_{12}O_{40}] \quad (3.74)$$

- The silicomolybdates/-tungstates $[SiMo_{12}O_{40}]^{4-}$/$[SiW_{12}O_{40}]^{4-}$ are prepared by acidification of MoO_4^{2-}/WO_4^{2-} using $NaSiO_3$ as a Si precursor (self-assembly), see eqn (3.72) and (3.73). POM acids can be isolated using the etherate method in an acidic environment. Using KCl, an ion exchange can be initiated, yielding the K^+ salt (see eqn (3.74)).[14,30]

3.6.1 Siliotungstate Lacunary Approaches

$$11Na_2WO_4 + Na_2SiO_3 + 16HCl \xrightarrow[\text{pH 5 to 6}]{100\,°C,\ 1\ hour} Na_8[SiW_{11}O_{39}] + 16NaCl + 8H_2O \quad (3.75)$$

$$Na_8[SiW_{11}O_{39}] + 2NaOH \xrightarrow{pH\ 9.1,\ 25\,°C,\ 16\ min} Na_8[SiW_{10}O_{36}] + Na_2WO_4 + H_2O \quad (3.76)$$

$$9Na_2WO_4 + Na_2SiO_3 + 11HCl \xrightarrow{PH\ 8} Na_9H[SiW_9O_{34}] + 11HCl + 5H_2O \quad (3.77)$$

$$2\ Na_9H[SiW_9O_{34}] + K_8[Nb_6O_{19}] + 12HCl + 4CsCl \xrightarrow{20\,°C} 2Cs_2K_4Na[SiW_9Nb_3O_{40}] + 16NaCl + 7H_2O \quad (3.78)$$

- Keggin lacunary-type anions of the silicotungstate with one, two and three vacancies are known. The mono lacunary-type anion $[SiW_{11}O_{39}]^{8-}$ is prepared from WO_4^{2-} with Na_2SiO_3 at pH values of 5 to 6 (self-assembly), see eqn (3.75). In the original procedure, those anions were directly precipitated as K^+ salts with KCl.[30] The lacunary-type POM is prepared here from the precursors and not from another lacunary-type species, meaning that the POM is prepared *de novo*. Basic solutions of $[SiW_{11}O_{39}]^{8-}$ at pH values of 9.1 (basification with Na_2CO_3) yield the two-fold lacunary-type species $[SiW_{10}O_{36}]^{8-}$ (see eqn (3.76)).[85,86]
- Keggin lacunary-type anions of the silicotungstate with three vacancies can be prepared by acidifying WO_4^{2-} solutions in

combination with Na_2SiO_3, yielding the anion $[SiW_9O_{34}]^{10-}$ at pH values of 8 (self-assembly), see eqn (3.77).[87]

- The vacant sites of $[SiW_9O_{34}]^{10-}$ can be filled successfully, *e.g.* with Nb(v) in a separate step (by adding $[Nb_6O_{19}]^{8-}$), yielding the Keggin-type TMSPOM $[SiNb_3W_9O_{40}]^{7-}$ (see eqn (3.78)).[87]

3.6.2 Ti(IV) Containing TMSPOMs

$$10Na_2WO_4 + Na_2HPO_4 + 2TiCl_4 + 7KCl \xrightarrow{100\ °C,\ pH\ 8.2} K_7[PTi_2W_{10}O_{40}] + 15NaCl + NaOH + 3Na_2O \quad (3.79)$$

$$11Na_2WO_4 + Co(CH_3COO)_2 + Ti(SO_4)_2 + 10CH_3OOH + 6KCl \xrightarrow{100\ °C,\ 1\ hour} K_6H_2[CoTiW_{11}O_{40}] + 6NaCl + 2Na_2SO_4 + 12CH_3OONa + 4H_2O \quad (3.80)$$

- TMSPOMs containing Ti(IV) as a foreign element are prepared in non-acidic media using titanium(IV) chloride $TiCl_4$. Here, sodium oxide Na_2O is presumably formed, which reacts directly with water, forming NaOH. This increases the pH value of the aqueous solution. The $[PTi_2W_{10}O_{40}]^{7-}$ anion is yielded in this way (see eqn (3.79)).[88]
- Another Keggin-type TMSPOM based on Co(II) as a heteroelement $[CoTiW_{11}O_{40}]^{8-}$ can be prepared starting from WO_4^{2-}, cobalt(II) acetate ($Co(CH_3COO)_2$), and titanyl sulfate $Ti(SO_4)_2$ (see eqn (3.80)).[89,90]

3.7 Purification of Polyoxometalates

After successful preparation of POM clusters, it is often necessary to purify them. In most approaches, unwanted alkali halogenides are formed as by-products because alkali containing precursor compounds are used and the pH value is adjusted with mineral acids like HCl. Stoichiometrically, more alkali halogenides are formed than POM molecules. Eqn (3.81) shows the formation of the alkali halogenide by-product. Here, 1 equivalent of POM, 6 equivalents of NaCl and 3 equivalents of water are formed.

$$Na_9[PMo_9O_{34}] + 3NaVO_3 + 6HCl \xrightarrow{20\ °C/reflux,\ pH\ 1} Na_6[PV_3Mo_9O_{40}] + 6NaCl + 3H_2O \quad (3.81)$$

Therefore, the need for purification becomes relevant if the aim is to isolate the POM as a pure substance from the reaction mixture. Possible purification approaches are as follows:

- A recrystallization of the POM material. This method is successful if the POM is soluble in warm water and crystallizes after the solution is cooled down. Isolation is then possible by filtration. A big disadvantage is that low product yields result, as the POM is often very soluble, even at low temperatures. Furthermore, the purification results can be low, as the alkali halogenide also crystallizes in some cases, meaning that a POM/alkali salt mixture is isolated.
- A precipitation of the POM with organic or inorganic cations. Adding cesium Cs^+ or tetrabutylammonium (TBA) cations to aqueous POM solutions results in the precipitation of a cation-modified POM compound, which can be isolated by filtration and washing several times with water. The POM is isolated with a purity of over 99% and an almost complete yield. However, the material can no longer be dissolved in water. In the case of Cs^+-modified POM salts, the material can no longer be dissolved non-destructively. The use of TBA-modified POMs is discussed separately in Chapter 4, Section 4.1.1.
- With the etherate method, the POM is isolated as a POM acid. Here, the POM is isolated as a pure substance and often with a high yield. The material is converted from an alkali salt (in the aqueous phase) into a POM acid (ether phase). However, the etherate method fails with highly negatively charged POM anions and cannot be used for every type of structure.
- The use of nanofiltration systems (diafiltration). A nanofiltration approach was established using a nanoporous membrane for purification of POMs.[10] The principle is based on the size exclusion principle, which is conducted as a dialysis variant. A big advantage is that the POM is isolated as an alkali salt with a high yield and purity.[10]

The nanofiltration desalination of POMs takes place on a nanoporous membrane, in which a feed solution (containing the alkali salt and the POM as a mixture) is passed through a system under high pressure. All alkali cations and chloride anions, with their comparatively small ionic size, can pass through the membrane and accumulate in the permeate fraction. POM anions are too large to cross the membrane and accumulate in the retentate fraction. After about four cycles of diafiltration, the alkali cation concentration reaches a value

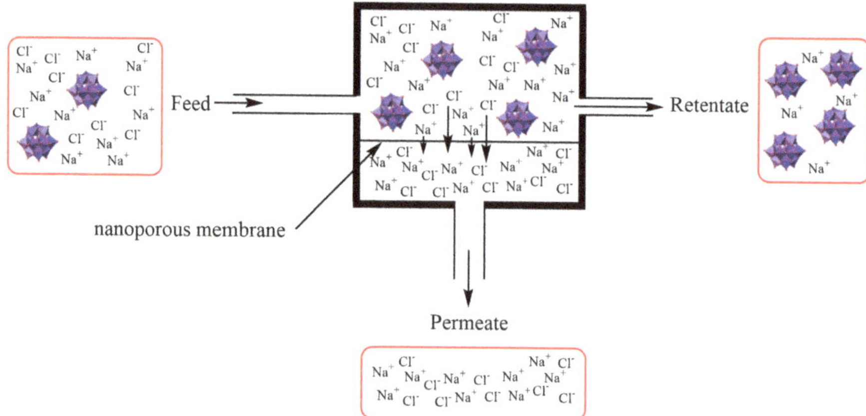

Figure 3.3 Principle of the nanofiltration/dialysis system. The feed fraction contains the POM/alkali salt mixture. Purified POM is enriched in the retentate and the alkali salt in the permeate fraction. Note: alkali cations are also required in a purified POM fraction in order to maintain electroneutrality with the POM anion.

required to ensure electroneutrality with the negatively charged POM anion. The principle of nanofiltration is visualized in Figure 3.3.[10]

Membrane processes are characterized by the so-called rejection in eqn (3.82):

$$R = 1 - \frac{c_{\text{permeate}}}{c_{\text{retentate}}} \tag{3.82}$$

where R is the rejection, and c_{permeate} and $c_{\text{retentate}}$ are the concentrations of the species in the permeate and retentate. The separation process is effective when the salt concentration is high in the permeate fraction and low in the retentate fraction. This means that the rejection for the salt is near to zero. Otherwise, the concentration of POM in the permeate must be close to zero, while the POM concentration in the retentate should not change. In this case, the rejection is near to 100%.[10,91–99]

The following parameters can be controlled in a nanofiltration process:[10,91–99]

- The pore size of the membrane. The larger the pores, the more large ions can pass through the membrane. Therefore, the pore size should not be too large, otherwise parts of the POM anion or dissociation fragments can cross the membrane, which reduces the rejection and the success of the process. On the other hand, the pore size must not be too small, otherwise the alkali cations and halogenide anions can no longer pass through the membrane.

- Feed concentration. The higher the concentration of the POM/alkali salt solution, the more ions are present on the membrane that are able to cross. This means that the separation process slows down. Care should be taken to ensure that the concentration is not too high. However, the selected concentration also depends on the size of the membrane system, especially the membrane surface area. With a larger membrane system, a higher feed concentration can be used.
- Flow rate. The flow rate describes the volume of feed solution passing through the system per minute. With a higher flow rate, a higher volume of the feed solution arrives at the membrane, which correlates with a higher number of ions that can pass through the membrane. The resulting effects are similar to those that arise when a higher concentration of the feed solution is used.
- Pressure. Pressure is one of the driving forces for the fluid to flow through the system. From a thermodynamic point of view, the higher the pressure, the greater the driving force for the ions to cross the membrane, see eqn (3.83). As the pressure increases, the separation speed of the process also increases. However, the pressure cannot be increased arbitrarily, as the membranes cannot withstand arbitrarily high pressures. Membranes usually have an upper pressure limit below which they may be operated. Exceeding this limit can lead to defects on the membrane.

$$\Delta\mu = \Delta\mu^0 + RT[\ln(a_{\text{retentate}}) - \ln(a_{\text{permeate}})] + V(p_{\text{retentate}} - p_{\text{permeate}}) \quad (3.83)$$

In eqn (3.83), $\Delta\mu$ is the overall change of the chemical potential, which needs to be negative if the ions should migrate from the feed/retentate into the permeate fraction. $\Delta\mu^0$ is the standard chemical potential, R is the ideal gas constant (8.314 J K^{-1} mol^{-1}), T is the temperature of the fluid, a is the activity (*e.g.* concentrations) of the ions in the retentate and permeate fractions, V is the volume of the fluid, and p is the pressure in the retentate and permeate fractions. Here, $p_{\text{rententate}}$ can be seen as the pressure with which the fluid is passed through the system and p_{permeate} as ambient pressure (1 bar). According to eqn (3.83) the migration of ions through the nanoporous membrane is driven by concentration and pressure. The driving force for ion migration is high, if the ion concentration difference between the retentate and permeate fraction is high and if the difference of pressure between the retentate and permeate fraction is high. This means that the thermodynamic driving force for ion migration from the retentate into the permeate fraction is higher (significantly more negative value of $\Delta\mu$) the more migrating ions are present in the retentate fraction and the greater the pressure with which the fluid is passed through the system.[10,91–99]

After several cycles of diafiltration, the feed fraction – an aqueous solution of the crude product from the synthesis, containing the alkali salt and the POM – is separated into a retentate fraction (purified POM) and a permeate fraction (isolated alkali salt, formed as a by-product in the synthesis). The salt and the POM can be obtained by removing the water. Although complete removal of the salt is theoretically only guaranteed after an infinite number of cycles, in practice it has been shown that the salt concentration drops below one per mille after four cycles.[10,91–99]

Abbreviations

POM	Polyoxometalate
TMSPOM	Transition metal-substituted polyoxometalate
pH	Negative decadic logarithm of the hydrogen ion concentration
HPA	Heterpolyanion/heteropolyacid
IPA	Isopolyanion/isopolyacid
AA	Acetic acid
RedOx	Reduction/oxidation

Acknowledgements

I would like to thank the publisher, the Royal Society of Chemistry, for the opportunity to write this book!

Recommended Reading

Please have a look into the following references: ref. (1) for the basics of element chemistry and (11) for the newly developed nanofiltration approach. For a broader application of the nanofiltration approach beyond polyoxometalate purification, see ref. (30). Note: ref. (1) is only available in German. Original literature on synthetic approaches involving the Wells–Dawson-type structure and Keggin-type lacunary derivatives can be found in ref. (21) to (24).

1. A. F. Holleman, E. und Nils Wiberg and G. Fischer, *Lehrbuch Der Anorganischen Chemie*, Berlin, New York, 2009.
2. S. Passadis, T. A. Kabanos, Y.-F. Song and H. N. Miras, *Inorganics*, 2018, **6**, 71.
3. V. F. Odyakov, E. G. Zhizhina and R. I. Maksimovskaya, *Appl. Catal., A*, 2008, **342**, 126–130.
4. V. F. Odyakov and E. G. Zhizhina, *Russ. J. Inorg. Chem.*, 2009, **54**, 361–367.
5. V. F. Odyakov and E. G. Zhizhina, *React. Kinet. Catal. Lett.*, 2008, **95**, 21–28.
6. C. Li, K. Yamaguchi and K. Suzuki, *Angew. Chem.*, 2021, **133**, 7036–7040.
7. C. Li, K. Yamaguchi and K. Suzuki, *Chem. Commun.*, 2021, **57**, 7882–7885.
8. C. Li, A. Jimbo, K. Yamaguchi and K. Suzuki, *Chem. Sci*, 2021, **12**, 1240–1244.
9. C. Li, N. Mizuno, K. Yamaguchi and K. Suzuki, *J. Am. Chem. Soc.*, 2019, **141**, 7687–7692.
10. S. Himeno, M. Hashimoto and T. Ueda, *Inorg. Chim. Acta*, 1999, **284**, 237–245.
11. J.-C. Raabe, T. Esser, F. Jameel, M. Stein, J. Albert and M. J. Poller, *Inorg. Chem. Front.*, 2023, **10**, 4854–4868.
12. J. F. Keggin, *Proc. R. Soc. London, Ser. A*, 1934, **144**, 75–100.
13. F. Kehrmann, *Z. Anorg. Chem.*, 1894, 7, 406–426.
14. B. Dawson, *Acta Crystallogr.*, 1953, **6**, 113–126.
15. A. F. Wells, Structural Inorganic Chemistry, Oxford, 1984.
16. J. S. Anderson, *Nature*, 1937, **140**, 1937.
17. H. T. Evans, *J. Am. Chem. Soc.*, 1948, **70**, 1291–1292.
18. I. Lindqvist, O. Hassel, M. Webb and M. Rottenberg, *Acta Chem. Scand.*, 1950, **4**, 1066–1074.
19. O. Linnenberg, A. Kondinski and K. Y. Monakhov, *Supramolecular Systems Chemistry*, 2017, pp. 39–66.
20. O. Nagano and Y. Sasaki, *Acta Crystallogr.*, 1979, **35**, 2387–2389.
21. I.-M. Mbomekalle, Y. W. Lu, B. Keita and L. Nadjo, *Inorg. Chem. Commun.*, 2004, 7, 86–90.
22. T. Moeller, *Inorganic Syntheses*, John Wiley & Sons, Inc, Hoboken, NJ, USA, 1990.
23. C. R. Graham and R. G. Finke, *Inorg. Chem.*, 2008, **47**, 3679–3686.
24. R. G. Finke, M. W. Droege and P. J. Domaille, *Inorg. Chem.*, 1987, **26**, 3886–3896.
25. D. R. Park, H. Kim, J. C. Jung, S. H. Lee and I. K. Song, *Catal. Commun.*, 2008, **9**, 293–298.

26. K. J. Schmidt, G. J. Schrobilgen and J. F. Sawyer, *Acta Crystallogr., Sect. C: Cryst. Struct. Commun.*, 1986, **42**, 1115–1118.
27. D. Drewes, E. M. Limanski and B. Krebs, *Eur. J. Inorg. Chem.*, 2004, **2004**, 4849–4853.
28. J.-C. Raabe, F. Jameel, M. Stein, J. Albert and M. J. Poller, *Dalton Trans.*, 2024, **53**, 454–466.
29. J.-C. Raabe, T. Esser, M. J. Poller and J. Albert, *Catal. Today*, 2024, **441**, 114899.
30. T. Esser, M. Huber, D. Voß and J. Albert, *Chem. Eng. Res. Des.*, 2022, **185**, 37–50.

References

1. A. F. Holleman, E. und N. Wiberg and G. Fischer, *Lehrbuch Der Anorganischen Chemie*, Berlin, New York, 2009.
2. V. F. Odyakov, E. G. Zhizhina and R. I. Maksimovskaya, *Appl. Catal., A*, 2008, **342**, 126–130.
3. V. F. Odyakov and E. G. Zhizhina, *Russ. J. Inorg. Chem.*, 2009, **54**, 361–367.
4. V. F. Odyakov and E. G. Zhizhina, *React. Kinet. Catal. Lett.*, 2008, **95**, 21–28.
5. C. Li, K. Yamaguchi and K. Suzuki, *Angew. Chem.*, 2021, **133**, 7036–7040.
6. C. Li, K. Yamaguchi and K. Suzuki, *Chem. Commun.*, 2021, **57**, 7882–7885.
7. C. Li, A. Jimbo, K. Yamaguchi and K. Suzuki, *Chem. Sci.*, 2021, **12**, 1240–1244.
8. C. Li, N. Mizuno, K. Yamaguchi and K. Suzuki, *J. Am. Chem. Soc.*, 2019, **141**, 7687–7692.
9. S. Himeno, M. Hashimoto and T. Ueda, *Inorg. Chim. Acta*, 1999, **284**, 237–245.
10. J.-C. Raabe, T. Esser, F. Jameel, M. Stein, J. Albert and M. J. Poller, *Inorg. Chem. Front.*, 2023, **10**, 4854–4868.
11. P. Pradeep, D.-L. Long, C. Streb and L. Cronin, *J. Am. Chem. Soc.*, 2008, **130**, 14946–14947.
12. R. S. Winter, J. M. Cameron and L. Cronin, *J. Am. Chem. Soc.*, 2014, **136**, 12753–12761.
13. S. Passadis, T. Kabanos, Y.-F. Song and H. Miras, *Inorganics*, 2018, **6**, 71.
14. J. D. H. Strickland, *J. Am. Chem. Soc.*, 1952, **74**, 862–867.
15. Y. Sun, J. Liu and E. Wang, *Inorg. Chim. Acta*, 1986, **117**, 23–26.
16. A. A. Mukhacheva, V. V. Volchek, V. V. Yanshole, N. B. Kompankov, A. L. Gushchin, E. Benassi, P. A. Abramov and M. N. Sokolov, *Inorg. Chem.*, 2020, **59**, 2116–2120.
17. M. J. Da Silva, N. A. Liberto, L. C. De Andrade Leles and U. A. Pereira, *J. Mol. Catal. A: Chem.*, 2016, **422**, 69–83.
18. S. Sheshmani, M. A. Fashapoyeh, M. Mirzaei, B. A. Rad, S. N. Ghortolmesh and M. Yousefi, *Indian J. Chem., Sect. A: Inorg., Bio-inorg., Phys., Theor. Anal. Chem.*, 2011, **50**, 1725–1729.
19. S. Himeno, M. Takamoto and T. Ueda, *Bull. Chem. Soc. Jpn.*, 2005, **78**, 1463–1468.
20. *Inorganic Syntheses*, ed. H. S. Booth, John Wiley & Sons, Inc, Hoboken, NJ, USA, 1939.
21. S. Himeno, M. Takamoto and T. Ueda, *J. Electroanal. Chem.*, 1999, **465**, 129–135.
22. S. M. Kulikov, O. M. Kulikova, R. I. Maksimovskaya and I. V. Kozhevnikov, *Bull. Acad. Sci. USSR, Div. Chem. Sci.*, 1990, **39**, 1763–1766.
23. J. A. Dias, S. C. L. Dias, E. Caliman, J. Bartis and L. Francesconi, *Inorganic Syntheses*, 2014, pp. 210–217.

24. *Inorganic Syntheses*, ed. G. S. Girolami and A. P. Sattelberger, Wiley, 2014, vol. 36.
25. J.-C. Raabe, M. J. Poller, D. Voß and J. Albert, *ChemSusChem*, 2023, **16**, e202300072.
26. J. M. Brégeault, M. Vennat, S. Laurent, J. Y. Piquemal, Y. Mahha, E. Briot, P. C. Bakala, A. Atlamsani and R. Thouvenot, *J. Mol. Catal. A: Chem.*, 2006, **250**, 177–189.
27. T. Ueda, K. Yamashita and A. Onda, *Appl. Catal., A*, 2014, **485**, 181–187.
28. J.-C. Raabe, J. Aceituno Cruz, J. Albert and M. J. Poller, *Inorganics*, 2023, **11**, 138.
29. I.-M. Mbomekalle, Y. W. Lu, B. Keita and L. Nadjo, *Inorg. Chem. Commun.*, 2004, **7**, 86–90.
30. T. Moeller, *Inorganic Syntheses*, John Wiley & Sons, Inc, Hoboken, NJ, USA, 1990.
31. C. R. Graham and R. G. Finke, *Inorg. Chem.*, 2008, **47**, 3679–3686.
32. R. G. Finke, M. W. Droege and P. J. Domaille, *Inorg. Chem.*, 1987, **26**, 3886–3896.
33. D. R. Park, H. Kim, J. C. Jung, S. H. Lee and I. K. Song, *Catal. Commun.*, 2008, **9**, 293–298.
34. H. Xu, Z. Bai, G. Wang, K. P. O'Halloran, L. Tan, H. Pang and H. Ma, *Microchim. Acta*, 2017, **184**, 4295–4303.
35. L. E. Briand, G. M. Valle and H. J. Thomas, *J. Mater. Chem.*, 2002, **12**, 299–304.
36. K. J. Schmidt, G. J. Schrobilgen and J. F. Sawyer, *Acta Crystallogr., Sect. C: Cryst. Struct. Commun.*, 1986, **42**, 1115–1118.
37. D. Drewes, E. M. Limanski and B. Krebs, *Eur. J. Inorg. Chem.*, 2004, **2004**, 4849–4853.
38. J.-C. Raabe, F. Jameel, M. Stein, J. Albert and M. J. Poller, *Dalton Trans.*, 2024, **53**, 454–466.
39. J. Albert, J. Mehler, J. Tucher, K. Kastner and C. Streb, *ChemistrySelect*, 2016, **1**, 2889–2894.
40. K. F. Jahr, J. Fuchs and R. Oberhauser, *Chem. Ber.*, 1968, **101**, 477–481.
41. J. Fuchs and K. F. Jahr, *Z. Naturforsch., B: J. Chem. Sci.*, 1968, **23**, 1380.
42. K.-H. Tytko and B. Schönfeld, *Z. Naturforsch., B: J. Chem. Sci.*, 1975, **30**, 471–484.
43. H. Avcı Özbek, *Chem. Pap.*, 2023, **77**, 5663–5669.
44. M. Nyman, T. M. Alam, F. Bonhomme, M. A. Rodriguez, C. S. Frazer and M. E. Welk, *J. Cluster Sci.*, 2006, **17**, 197–219.
45. M. Dabbabi and M. Boyer, *J. Inorg. Nucl. Chem.*, 1976, **38**, 1011–1014.
46. J. Fuchs, K. F. Jahr and G. Heller, *Chem. Ber.*, 1963, **96**, 2472–2484.
47. V. G. Maiorov, A. I. Nikolaev, V. K. Kopkov, V. Y. Kuznetsov and N. L. Mikhailova, *Russ. J. Appl. Chem.*, 2011, **84**, 1137–1140.
48. L. C. W. Baker, G. A. Gallagher and T. P. McCutcheon, *J. Am. Chem. Soc.*, 1953, **75**, 2493–2495.
49. L. C. W. Baker and T. J. R. Weakley, *J. Inorg. Nucl. Chem.*, 1966, **28**, 447–454.
50. R. D. Hall, *J. Am. Chem. Soc.*, 1907, **29**, 692–714.
51. L. Y. Feng, Y. H. Wang, Y. J. Qi, C. W. Hu, Y. Xu and E. B. Wang, *J. Mol. Struct.*, 2003, **645**, 231–234.
52. B. Hedman, Multicomponent polyanions. 16. The molecular and crystal structure of $Na_6Mo_5P_2O_{23}(H_2O)_{14}$, a compound containing sodium-coordinated pentamolybdodiphosphate anions, *Acta Crystallogr., Sect. B*, 1977, **B33**, 3083–3090.
53. C. Dey, *Coord. Chem. Rev.*, 2024, **510**, 215847.
54. I. Creaser, M. C. Heckel, R. J. Neitz and M. T. Pope, *Inorg. Chem.*, 1993, **32**, 1573–1578.
55. R. Kato, A. Kobayashi and Y. Sasaki, *J. Am. Chem. Soc.*, 1980, **102**, 6571–6572.
56. J. Fuchs, *Z. Naturforsch., B: J. Chem. Sci.*, 1973, **28**, 389–404.
57. M. Filowitz, R. K. C. Ho, W. G. Klemperer and W. Shum, *Inorg. Chem.*, 1979, **18**, 93–103.
58. A. Chemseddine, C. Sanchez, J. Livage, J. P. Launay and M. Fournier, *Inorg. Chem.*, 1984, **23**, 2609–2613.

59. R. D. Peacock and T. J. R. Weakley, *J. Chem. Soc. A*, 1971, 1836.
60. T. R. Zhang, R. Lu, H. Yu Zhang, P. C. Xue, W. Feng, X. L. Liu, B. Zhao, Y. Y. Zhao, T. J. Li and J. N. Yao, *J. Mater. Chem.*, 2003, **13**, 580–584.
61. Y. Guo, D. Li, C. Hu, Y. Wang, E. Wang, Y. Zhou and S. Feng, *Appl. Catal., B*, 2001, **30**, 337–349.
62. A. J. Bridgeman, *J. Phys. Chem. A*, 2002, **106**, 12151–12160.
63. B. Krebs and I. Paulat-Böschen, *Acta Crystallogr., Sect. B: Struct. Sci., Cryst. Eng. Mater.*, 1982, **38**, 1710–1718.
64. A. Müller, S. K. Das, V. P. Fedin, E. Krickemeyer, C. Beugholt, H. Bögge, M. Schmidtmann and B. Hauptfleisch, *Z. Anorg. Allg. Chem.*, 1999, **625**, 1187–1192.
65. C. Marchal-Roch, E. Ayrault, L. Lisnard, J. Marrot, F.-X. Liu and F. Sécheresse, *J. Cluster Sci.*, 2006, **17**, 283–290.
66. P. J. Domaille and G. Watunya, Synthesis and tungsten-183 NMR characterization of vanadium-substituted polyoxometalates based on B-type tungstophosphate PW_9O_{349}-precursors, *Inorg. Chem.*, 1986, **25**(8), 1239–1242.
67. J. H. Kyle, *J. Chem. Soc., Dalton Trans.*, 1983, 2609–2612.
68. L. A. Combs-Walker and C. L. Hill, *Inorg. Chem.*, 1991, **30**, 4016–4026.
69. A. Patel and S. Pathan, *J. Coord. Chem.*, 2012, **65**, 3122–3132.
70. A. J. Gaunt, I. May, M. J. Sarsfield, D. Collison, M. Helliwell and I. S. Denniss, *Dalton Trans.*, 2003, **3**, 2767–2771.
71. R. Massart, R. Contant, J. M. Fruchart, J. P. Ciabrini and M. Fournier, *Inorg. Chem.*, 1977, **16**, 2916–2921.
72. *Inorganic Syntheses*, ed. M. Y. Darensbourg, Wiley, Hoboken, NJ, USA, 1998.
73. D. R. Park, J. H. Choi, S. Park and I. K. Song, *Appl. Catal., A*, 2011, **394**, 201–208.
74. D. K. Lyon, W. K. Miller, T. Novet, P. J. Domaille, E. Evitt, D. C. Johnson and R. G. Finke, *J. Am. Chem. Soc.*, 1991, **113**, 7209–7221.
75. G. A. Tsigdinos and C. J. Hallada, *Inorg. Chem.*, 1968, **7**, 437–441.
76. J. H. Grate, *J. Mol. Catal. A: Chem.*, 1996, **114**, 93–101.
77. E. G. Zhizhina, V. F. Odyakov and M. V. Simonova, *Kinet. Catal.*, 2008, **49**, 773–781.
78. E. G. Zhizhina and V. F. Odyakov, *React. Kinet. Catal. Lett.*, 2008, **95**, 301–312.
79. T. Ueda, Y. Nishimoto, R. Saito, M. Ohnishi and J. Nambu, *Inorganics*, 2015, **3**, 355–369.
80. M. Abbessi, R. Contant, R. Thouvenot and G. Hervé, *Inorg. Chem.*, 1991, **30**, 1695–1702.
81. L. E. Briand, H. J. Thomas and G. T. Baronetti, *Appl. Catal., A*, 2000, **201**, 191–202.
82. L. H. Bi, E. B. Wang, J. Peng, R. D. Huang, L. Xu and C. W. Hu, *Inorg. Chem.*, 2000, **39**, 671–679.
83. J. H. Choi, D. R. Park, S. Park and I. K. Song, *Korean J. Chem. Eng.*, 2011, **28**, 2137–2141.
84. L. M. Sanchez, Á. G. Sathicq, G. T. Baronetti, H. J. Thomas and G. P. Romanelli, *Catal. Lett.*, 2014, **144**, 172–180.
85. G. M. Maksimov, R. I. Maksimovskaya and O. A. Kholdeeva, *Russ. J. Inorg. Chem.*, 2004, **49**, 1436–1441.
86. J. Canny, A. Teze, R. Thouvenot and G. Herve, *Inorg. Chem.*, 1986, **25**, 2114–2119.
87. Y. Qi, Y. Xiang, J. Wang, Y. Qi, J. Li, J. Niu and J. Zhong, *Antiviral Res.*, 2013, **100**, 392–398.
88. P. J. Domaille and W. H. Knoth, *Inorg. Chem.*, 1983, **22**, 818–822.
89. Y. Chen and J. Liu, *Synth. React. Inorg. Met.-Org. Chem.*, 1997, **27**, 239–250.
90. H. Kraus, H. Stephan, A. Röllich, Z. Matějka and G. Reck, *Acta Crystallogr., Sect. E: Struct. Rep. Online*, 2005, **61**, i35–i37.
91. T. Esser, M. Huber, D. Voß and J. Albert, *Chem. Eng. Res. Des.*, 2022, **185**, 37–50.
92. K. Sarkar, A. K. SenGupta and P. Prakash, *Sci. Technol.*, 2010, **44**, 1161–1166.
93. G. Stell and C. G. Joslin, *Biophys. J.*, 1986, **50**, 855–859.

94. R. Epsztein, E. Shaulsky, N. Dizge, D. M. Warsinger and M. Elimelech, *Environ. Sci. Technol.*, 2018, **52**, 4108–4116.
95. N. N. R. Ahmad, W. L. Ang, Y. H. Teow, A. W. Mohammad and N. Hilal, *J. Water Process Eng.*, 2022, **45**, 102478.
96. B. Tansel, J. Sager, T. Rector, J. Garland, R. F. Strayer, L. Levine, M. Roberts, M. Hummerick and J. Bauer, *Sep. Purif. Technol.*, 2006, **51**, 40–47.
97. N. S. Suhalim, N. Kasim, E. Mahmoudi, I. J. Shamsudin, A. W. Mohammad, F. Mohamed Zuki and N. L.-A. Jamari, *Nanomaterials*, 2022, **12**, 437.
98. D. Vezzani and S. Bandini, *Desalination*, 2002, **149**, 477–483.
99. M. Qasim, N. N. Darwish, S. Mhiyo, N. A. Darwish and N. Hilal, *Desalination*, 2018, **443**, 143–164.

4 Properties of Polyoxometalates

4.1 Chemical Properties

An element from the periodic table can occupy different positions in a POM cluster (Figure 4.1). The most important positions are:

- Cation position. Controls the solubility properties of a POM cluster.
- Framework element position. Controls the RedOx activity.
- Foreign element position. Controls the RedOx properties and has a big impact on the overall charge.
- Heteroelement position. Only relevant in HPA structure types. Controls the formation of the structure type.
- Ligand position. Mostly oxygen (hence the name "oxometalates"), but there are also some other elements that can act as ligands for the metals. The ligand controls the stability of a POM cluster. In general, a distinction must be made between terminal $M{=}O_t$ and bridging M–O–M oxo ligands.
- Hydration water. POMs synthesized in aqueous media are always isolated as multihydrates.

In this subchapter, the cation, heteroelement, ligand and hydration water positions are discussed. The roles of the framework elements and foreign elements are discussed in Chapter 5.

RSC Foundations No. 3
Polyoxometalate Chemistry
By Jan-Christian Raabe

Published by the Royal Society of Chemistry, www.rsc.org

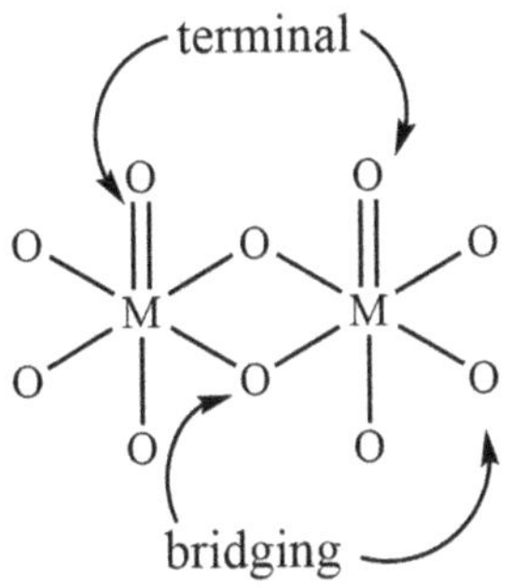

Figure 4.1 Different types of oxo ligands found in POMs.

4.1.1 The Cation Position

POMs with different cations can be synthesized, *e.g.* inorganic cations (such as protons, *i.e.* POM acids), alkali or transition element cations, and organic cations.[1–8]

4.1.1.1 General Tendencies

POM alkali salts show very high water solubility, especially for the cations Li^+, Na^+ and K^+, but show no solubility for Rb^+ and Cs^+ salts. Cs^+ salts of POMs are mostly completely insoluble in water or any solvent.[9–11] Dissolution is only possible by destroying the POM cluster, *e.g.* with HF digestion, forming metal fluorides. The solubility of POM compounds in aqueous media decreases from H^+ to Li^+ and finally to Cs^+. This can be explained by the so-called **H**ard and **S**oft **A**cids and **B**ases (HSAB) concept. POM anions are considered as "soft" bases due to the delocalization of the anionic charge across the whole cluster, while the cations H^+, Li^+, Na^+ and K^+ are more likely to be classified as "hard" acids. Cs^+ cations are "soft" cations, so a POM anion forms a stable interaction with a Cs^+ cation, which means the crystal lattice cannot be broken up by water molecules.[12–15] Since Cs^+ is significantly larger than H^+ or Li^+, its positive charge is distributed over a much larger ion. The ionic potential of the Cs^+ cation is therefore significantly lower than that of Li^+, meaning that the hydrate shell of Cs^+ can be removed more easily than that of Li^+. This makes it easier to form the Cs^+ POM salt from aqueous solution.[16] A similar effect is observed when the POM anion is precipitated with organic cations, such as TBA^+. However, the TBA^+ cation shows a high solubility in organic solvents, such as acetonitrile, so that the

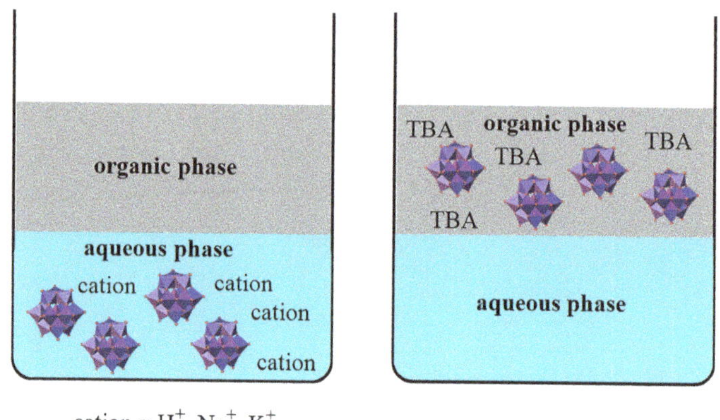

$$\text{cation}_n\text{POM} + n\ \text{CsCl} \longrightarrow \text{Cs}_n\text{POM} + n\ \text{cationCl}$$

$$\text{cation}_n\text{POM} + n\ \text{TBABr} \longrightarrow \text{TBA}_n\text{POM} + n\ \text{cationBr}$$

Figure 4.2 Cation modification of POMs. Precipitating of POMs with Cs^+ cations results in an insoluble residue and precipitating POMs with TBA^+ cations forms a material that is soluble in organic solvents, *e.g.* acetonitrile.

entire POM cluster dissolves in polar organic solvents. This means that the POM material can be transferred from the aqueous to the organic phase and so organic chemistry with POMs is possible (see Figure 4.2).[8]

4.1.1.2 Cations in Preyssler-type POMs

Preyssler-type anions can complex simple inorganic cations with their large internal cavity. The Preyssler-type anion $[NaP_5W_{30}O_{110}]^{14-}$ was investigated to study this behavior.[17]

This complex behaves like a separate anion, which was isolated as NH_4^+ salt $(NH_4)_{14}[NaP_5W_{30}O_{110}]$ and the Na^+ cation was obviously not exchanged by the NH_4^+ cations, as shown in Figure 4.3.[16,17] Similar complexes $[M^nP_5W_{30}O_{110}]^{(15-n)-}$ were found (n oxidation state) for $M = Ca^{2+}$, Nd^{3+}, Sm^{3+}, Cd^{2+}, Sn^{2+}, Hg^{2+} and more.[18,19] Complexation of higher-valent cations is much more difficult and the reaction solution must be heated for a successful complexation. More energy has to be applied for cations with higher charges in order to remove them from their water hydration shells.[16,19]

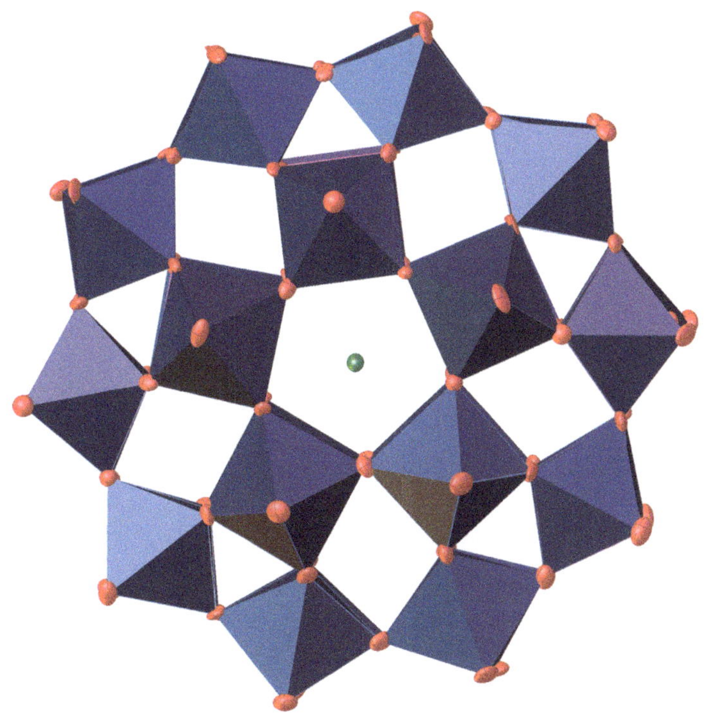

Figure 4.3 The Preyssler-type structure $[MP_5W_{30}O_{110}]^{15-}$ with an additional complexed cation M (green). Color code: purple – P, blue – W, red – O and green – cation M. The data were used from the Cambridge Crystallographic Data Centre and Fachinformations-zentrum Karlsruhe Access Structures service database (deposition number: 1014312).

4.1.1.3 POMs as Ionic Liquids

Using special organic cations (as shown in Figure 4.4) in combination with POM anions, it is possible to prepare POM ionic liquids (POM IL).[20–23]

ILs are classified as low-melting salts that form liquids below the boiling point of water. Every IL consists of only cations and anions. There are many synonyms used for ILs, *e.g.* molten salt, ionic fluid, liquid organic salt, fused salt or neoteric solvent. In the field of green chemistry, where chemical processes are designed to reduce the use of hazardous substances, it is desirable to avoid volatile organic solvents. Suitable solvents could be water, supercritical fluids or ILs.[24]

With cations, such as those shown in Figure 4.4, and polyanions, such as $[W_6O_{19}]^{2-}$, $[PW_{12}O_{40}]^{3-}$, $[SiW_{12}O_{40}]^{4-}$ or $[PMo_{12}O_{40}]^{3-}$, different POM ILs can be prepared.[20–22]

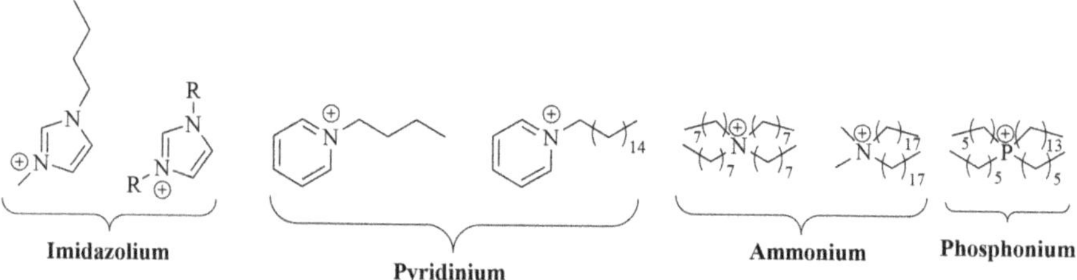

Figure 4.4 Different organic cations for the modification of POMs in order to prepare POM ILs. For example, imidazolium, pyridinium, ammonium and phosphonium cations can be used.

4.1.1.4 POMs with Main-group Element Cations

Different main-group element cations were used to modify the Keggin-type POM anion $[PW_{12}O_{40}]^{3-}$, *e.g.* Al(III), Ga(III), In(III) and Bi(III). A preparation of those salts was achieved by a simple ion-exchange reaction, in which the main-group element nitrate was converted with the POM acid $H_3[PW_{12}O_{40}]$.[25–28] Salt $Bi[PW_{12}O_{40}]$ was identified as an insoluble, colorless compound.[28]

4.1.1.5 POMs with Transition Element Cations

In addition to the use of alkali or organic cations, POMs with transition metal cations have also been described, like $Fe[PW_{12}O_{40}]$ or $Ag_3[PW_{12}O_{40}]$, which can be prepared by precipitation of the POM by adding transition element salts.[29–31]

4.1.1.6 Sandwich-type POMs

Sandwich-type POMs are at least two lacunary-type clusters that complex inorganic cations. The cations are found in a sandwich-like arrangement between the two clusters.

Examples are the Keggin/Wells–Dawson lacunary-type clusters $[Co_4(H_2O)_2(PW_9O_{34})_2]^{10-}$, $[Zn_4(H_2O)_2(PW_9O_{34})_2]^{10-}$, $[Mn_4(H_2O)_2(P_2W_{15}O_{56})_2]^{16-}$ and $[Cu_4(H_2O)_2(P_2W_{15}O_{56})_2]^{16-}$, which formally consist of two Keggin/Wells–Dawson lacunary-type anions that coordinate to four Co^{2+}, Zn^{2+}, Mn^{2+} or Cu^{2+} cations with their vacant sides. Formally, each transition metal cation is octahedrally surrounded by six other oxo ligands. There are two cations that have no coordination partner on the outside. Additional water ligands or a second type of ligand (*e.g.* organic amines) can coordinate here. The four cations are

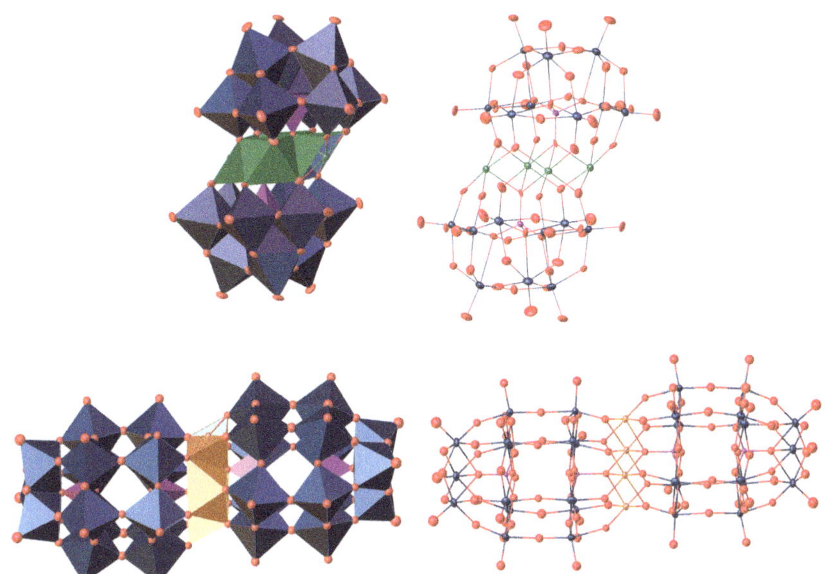

Figure 4.5 Sandwich-type POMs with the general molecular formula $[X_2M_{18}C_4(H_2O)_2O_{68}]^{n-}$ or $[C_4(H_2O)_2(XM_9O_{34})2]^{n-}$ (Keggin lacunary-type, top) and $[C_4(H_2O)_2(P_2W_{15}O_{56})_2]^{16-}$ (Wells-Dawson lacunary-type, bottom). C are additional cations (*e.g.* Co^{2+}, Zn^{2+}, Mn^{2+} or Cu^{2+}) that are coordinated by two lacunary-type anions. Polyhedron (left) and atomistic representation (right). Color code: purple - X, blue - M, red - O and green/orange - cation C. Coordinating water ligands are not shown. The data were used from the Cambridge Crystallographic Data Centre and Fachinformationszentrum Karlsruhe Access Structures service database (deposition number: 871204, Keggin lacunary-type and 753710, Wells-Dawson lacunary-type).

arranged in a typical parallelogram shape. Typical three-dimensional representations of those cluster-types are shown in Figure 4.5.[32,33]

A very similar cluster has been reported, in which Na^+ cations are complexed by two Keggin lacunary-type anions of an arsenic-containing POM with the molecular formula $[Na_6(H_2O)_x(AsW_9O_{34})_2]^{12-}$.[34] The POM fragment $[AsW_9O_{34}]^{9-}$ is called arsenotungstate. Here, six Na^+ cations are found in a hexagonal arrangement. Each Na^+ cation is octahedrally coordinated by six oxo ligands, with up to twelve water ligands that could be coordinated (six in the interior and six outside of the cluster). The three-dimensional structure of this cluster is shown in Figure 4.6.[34]

Another difference between the two clusters shown in Figure 4.5 and 4.6 is found in the orientation of the central XO_4 tetrahedron. The XO_4 tetrahedron in the compound $[C_6(H_2O)_x(XM_9O_{34})_2]^{n-}$ (Figure 4.6) is oriented in a typical Keggin lacunary-type arrangement (Figure 1.7). In the compound $[Co_4(H_2O)_2(PW_9O_{34})_2]^{n-}$ (Figure 4.5) the XO_4

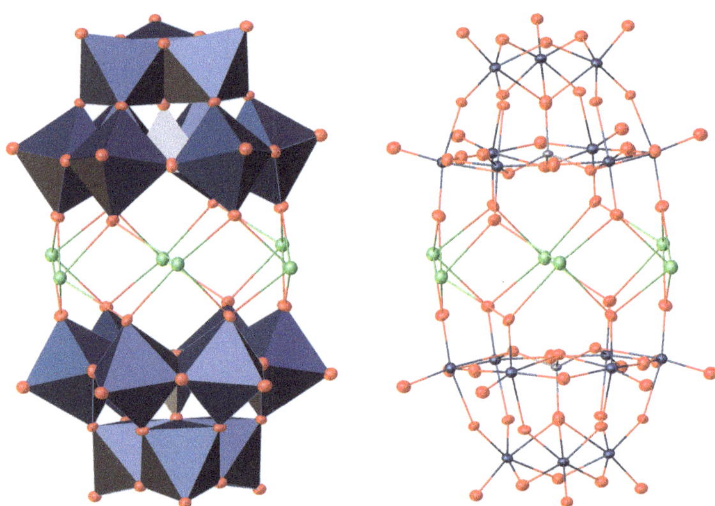

Figure 4.6 Sandwich-type POMs with the general molecular formula $[M_6(H_2O)_xX_2M_{18}O_{68}]^{n-}$ or $[M_6(H_2O)_x(PW_9O_{34})_2]^{n-}$. M are additional cations (*e.g.* Na^+), that are coordinated by two Keggin lacunary-type anions. Polyhedron (left) and atomistic representation (right). Color code: gray – X, blue – M, red – O and green – cation M. Coordinating water ligands are not shown. The data were used from the Cambridge Crystallographic Data Centre and Fachinformationszentrum Karlsruhe Access Structures service database (deposition number: 2293851).

tetrahedron protrudes with one of the four oxo ligands into the vacant sides and coordinates to three of the four cations. This means that the metalate framework is held together by only three of the four oxo ligands of XO_4. Therefore, the fourth oxo ligand does not coordinate to the metalate framework.

In general, sandwich-type POMs of Wells–Dawson- and Keggin-type can be prepared as follows:

- Sandwich-type clusters of the Wells–Dawson lacunary-type anion $[P_2W_{15}O_{56}]^{12-}$, like $Na_{16}[Mn_4(H_2O)_2(P_2W_{15}O_{56})_2]$, are prepared by adding simple transition element chlorides (like $MnCl_2$) to an aqueous solution of the lacunary-type species $[P_2W_{15}O_{56}]^{12-}$ (see eqn (4.1)).[35]
- Anions $[Co_4(H_2O)_2(PW_9O_{34})_2]^{10-}$ and $[Zn_4(H_2O)_2(PW_9O_{34})_2]^{10-}$ are prepared *de novo* starting from WO_4^{2-}, PO_4^{3-} and salts of the cations Co(II) or Zn(II) at a pH value of 4 and at 160 °C (see eqn (4.2) and (4.3)). If two amine ligands are required to complex to the cations instead of water ligands, the amine must be added in the last step.[32]

- The anion $[Na_6(H_2O)_x(AsW_9O_{34})_2]^{12-}$ is prepared *de novo* by WO_4^{2-} and arsenic acid H_3AsO_4. Both compounds are acidified to pH values between 5 and 1, yielding the Na^+ salt with the stoichiometric composition of $Na_{12}[Na_6(H_2O)_x(AsW_9O_{34})_2]$ (see eqn (4.4)).[34]
- Keggin lacunary-type clusters like $[SbW_9O_{33}]^{9-}$ are prepared in non-acidic media, starting from WO_4^{2-} and antimony(III) oxide Sb_2O_3 (see eqn (4.5)). It is likely that formally Na_2O is formed, which directly reacts with water to form NaOH. NaOH formation is therefore responsible for the observed increasing pH value of the reaction solution.[36] The formation of a complete, intact and spherical structure is suppressed by the free electron pair on Sb(III) and by the enormous ionic size. The formation of open structural motifs is thermodynamically favored due to the structural tension in spherical motifs caused by bigger atoms. Due to the free electron pair of Sb(III), no fourth oxygen atom can bind to the heteroelement, so Sb(III) is found in a trigonal-pyramidal coordination environment in the Keggin lacunary-type structure (one oxygen atom is missing—there are 33 oxygen atoms instead of 34 for normal Keggin lacunary-type clusters).[34,36]
- The lacunary-type anion $[SbW_9O_{33}]^{9-}$ is also able to form sandwich-type dimeric clusters. For example, VO^{2+} cations can be complexed by adding vanadyl sulfate $VOSO_4$ to aqueous solutions of the lacunary-type species. The complex anion $[(V^{IV}O)_3(SbW_9O_{33})_2]^{12-}$ can be successfully isolated (see eqn (4.6)).[37]
- Dimers of Keggin lacunary-type POMs containing Zn(II) as heteroelement $[ZnW_9O_{34}]^{12-}$ are able to complex a W(VI) and three Zn(II) cations by acidification of aqueous WO_4^{2-} and $Zn(NO_3)_2$ solutions, yielding anions with the molecular stoichiometry of $[WZn_3(H_2O)_2(ZnW_9O_{34})_2]^{12-}$ (see eqn (4.7)).[38]

A special case of sandwich-type POMs is the so-called Krebs-type structure (named after Bernt Krebs), with the general molecular formula of $[(M(H_2O)_3)_2(WO_2)_2(XW_9O_{33})_2]^{n-}$ or $[(M(H_2O)_3)_2(W\{H_2O\}_2)_2(XW_9O_{33})_2]^{n-}$.[39] This type of structure can be understood as a separate structure type or as a simple sandwich-type cluster consisting of two Keggin lacunary-type clusters $[XW_9O_{33}]^{m-}$ (comparable to the anion $[SbW_9O_{33}]^{9-}$) that complex two transition element M(II) cations and two W(IV) cations. Three water ligands coordinate to each of the M(II) cations. However, the literature does not clearly differentiate whether two water ligands or two terminal oxo ligands coordinate to the W(VI) cations. To compensate for the high cationic charges of

W(VI), it is more likely that two terminal oxo ligands coordinate to the W(VI) cations. This results in $[WO_2]^{2+}$ fragments that are coordinated by the lacunar oxo ligands. A $[WO_2]^{2+}$ fragment can also be partially protonated, forming a more cationic charged $[W(OH)_2]^{4+}$ fragment (with hydroxido ligands). The two Keggin lacunary-type monomers can contain Sb(III), Bi(III) or Te(VI) as heteroelements and can coordinate to two $[WO_2]^{2+}$/$[W(OH)_2]^{4+}$ units, as well as to two M(II) cations. Comparable to the $[SbW_9O_{33}]^{9-}$ anion, the heteroelement is trigonal-pyramidal coordinated by three oxo ligands, since both Bi(III) and Sb(III) have a free electron pair. The first reported anions were the Mn(II) complexes $[(Mn^{II}(H_2O)_3)_2(WO_2)_2(BiW_9O_{33})_2]^{10-}$(see eqn (4.8)) and $[(Mn^{II}(H_2O)_3)_2(Mn^{II}(H_2O)_2)_2(TeW_9O_{33})_2]^{8-}$ (see eqn (4.9)).[39] Here, the Te(IV) precursor is prepared by dissolving tellurium(IV) oxide TeO_2 in water. Tellurous acid H_2TeO_3 is formed, which can get neutralized by NaOH, yielding tellurate(IV) $TeO_3{}^{2-}$ anions (see eqn (4.10)).[40] According to this concept, the $[(Mn^{II}(H_2O)_3)_2(Mn^{II}(H_2O)_2)_2(TeW_9O_{33})_2]^{8-}$ cluster is not a Krebs-type structure, but it was mentioned in the original publication by Bernt Krebs.[39]

Sandwich-type Clusters of the Wells–Dawson-type Structure

$$2Na_{12}[P_2W_{15}O_{56}] + 4MnCl_2 + 2H_2O \xrightarrow{20\ ^\circ C} Na_{16}[Mn_4(H_2O)_2(P_2W_{15}O_{56})_2] + 8NaCl \quad (4.1)$$

Sandwich-type Clusters of the Keggin-type Structure

$$18Na_2WO_4 + 2NaH_2PO_4 + 4CoCl_2 + 24HCl + 2H_2O \xrightarrow[\text{pH 4}]{160\ ^\circ C,\ 4\ \text{days}} Na_{10}\left[Co_4(H_2O)_2(PW_9O_{34})_2\right] + 28NaCl + 12H_2O \quad (4.2)$$

$$18Na_2WO_4 + 2NaH_2PO_4 + 4ZnCl_2 + 24HCl + 2H_2O \xrightarrow[\text{pH 4}]{160\ ^\circ C,\ 4\ \text{days}} Na_{10}\left[Zn_4(H_2O)_2(PW_9O_{34})_2\right] + 28NaCl + 12H_2O \quad (4.3)$$

$$18Na_2WO_4 + 2H_3AsO_4 + xH_2O + 18HCl \xrightarrow{20\ ^\circ C,\ \text{pH 5–1}} Na_{12}[Na_6(H_2O)_x(AsW_9O_{34})_2] + 18NaCl + 12H_2O \quad (4.4)$$

$$18Na_2WO_4 + Sb_2O_3 \xrightarrow{20\ ^\circ C,\ \text{non-acidic}} 2Na_9[SbW_9O_{33}] + 9Na_2O \quad (4.5)$$

$$2Na_9[SbW_9O_{33}] + 3VOSO_4 + 11KCl + CH_3COOH \xrightarrow{60\text{–}70\ ^\circ C,\ 1\ hour} K_{11}H[(VO)_3(SbW_9O_{33})_2] + 11NaCl + 3Na_2SO_4 + CH_3COONa \quad (4.6)$$

$$19Na_2WO_4 + 5Zn(NO_3)_2 + 16HNO_3 + 2H_2O \xrightarrow{80\text{–}95\ ^\circ C,\ 2\text{–}3\ hours} Na_{12}[WZn_3(H_2O)_2(ZnW_9O_{34})_2] + 26NaNO_3 + 8H_2O \quad (4.7)$$

$$20Na_2WO_4 + 2BiONO_3 + 2MnCl_2 + 24HNO_3 + 4NH_4NO_3 + 6H_2O \xrightarrow[75\ ^\circ C]{pH\ 5\text{–}6,\ 1\ hour} Na_6(NH_4)_4[(Mn(H_2O)_3)_2(WO_2)_2(BiW_9O_{33})_2] + 4NaCl + 30NaNO_3 + 12H_2O \quad (4.8)$$

$$18Na_2WO_4 + 2Na_2TeO_3 + 4MnCl_2 + 24HNO_3 + 10H_2O \xrightarrow{pH\ 3,\ 1\ hour\ 80\ ^\circ C} Na_8[(Mn(H_2O)_3)_2(Mn(H_2O)_2)_2(TeW_9O_{33})_2] + 8NaCl + 24NaNO_3 + 12H_2O \quad (4.9)$$

$$TeO_2 + H_2O \rightleftharpoons H_2TeO_3$$
$$H_2TeO_3 + 2NaOH \rightleftharpoons Na_2TeO_3 + 2H_2O \quad (4.10)$$

$$20Na_2WO_4 + 2Bi(NO_3)_3 + 2CoCl_2 + 26HCl + imi + 7H_2O \xrightarrow{pH\ 6.8,\ 2\ hours\ 80\ ^\circ C} Na_4(H_2imi)_2[Co_2(H_2O)_6\{W(OH)_2\}_2(BiW_9O_{33})_2] + 30NaCl + 6NaNO_3 + 13H_2O \quad (4.11)$$

$$Na_7[PW_{11}O_{39}] + PdCl_2 + 5KCl \xrightarrow{90\ ^\circ C,\ 1\ hour} K_5[PPdW_{11}O_{39}] + 7NaCl \quad (4.12)$$

$$Na_7[XM_{11}O_{39}] + M'Cl_2 + 5KCl \rightarrow K_5[XM'M_{11}O_{39}] + 7NaCl \quad (4.13)$$

Today, more representatives are known, *e.g.* the Co(II) complex $[Co_2(H_2O)_6\{W(OH)_2\}_2(BiW_9O_{33})_2]^{6-}$, which was isolated as protonated imidazole (imi/H$_2$imi) salt (see eqn (4.11)). The three-dimensional structure of a typical Krebs-type cluster is illustrated in Figure 4.7.[41]

The anion $[(WZn_3(H_2O)_2(ZnW_9O_{34})_2]^{12-}$ can also be interpreted as a special case of the Krebs-type structure (see eqn (4.2)). However, the W(VI) cation is fully integrated into the lacunary-type cluster, formally forming a $[ZnW_{10}O_{34}]^{6-}$ cluster, which complexes three Zn(II) cations together with a $[ZnW_9O_{34}]^{12-}$ unit.

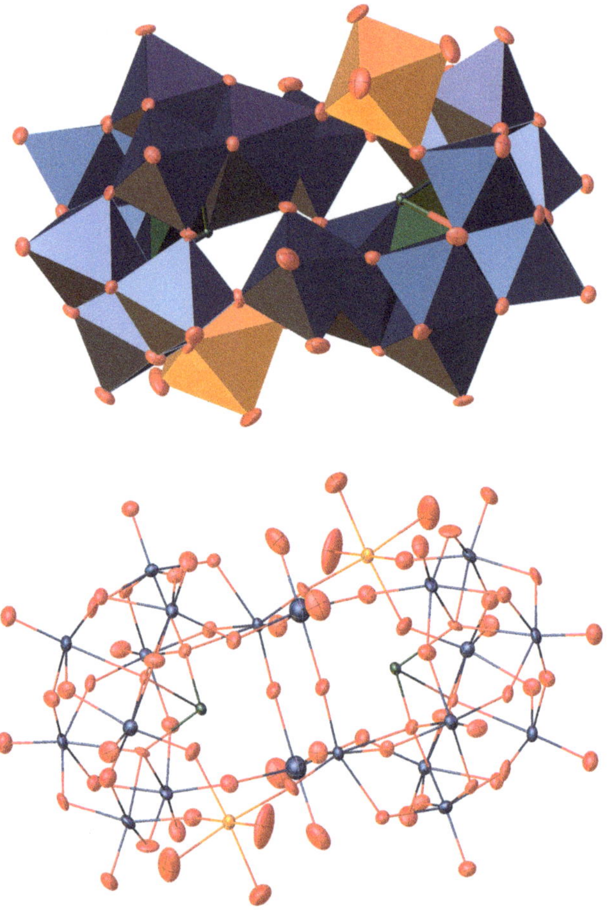

Figure 4.7 Krebs-type structure as a special case of a sandwich-type POM in the polyhedron (top) and atomistic (bottom) model. Color code: green – X (heteroelement: Bi(III), Sb(III) or Te(IV)), blue – W, red – O and orange foreign transition element cation M(II) (*e.g.* Co(II) or Mn(II)). The data were used from the Cambridge Crystallographic Data Centre and Fachinformationszentrum Karlsruhe Access Structures service database (deposition number: 730764).

Mono Keggin lacunary-type POMs $[XM_{11}O_{39}]^{n-}$ can complex simple transition element cations, forming compounds like $[XM'M_{11}O_{39}]^{m-}$, as shown in Figure 4.8. Prominent examples are anions like $[PPdW_{11}O_{39}]^{5-}$, $[PCoW_{11}O_{39}]^{5-}$, $[PCuW_{11}O_{39}]^{5-}$ or $[SiRuW_{11}O_{39}]^{5-}$. Those complexes are prepared by adding the transition element halide to aqueous solutions of the Keggin lacunary-type species (see eqn (4.12) and eqn (4.13)). The driving force of this reaction is the reduction of negative charge from the lacunary-type species. A complexed transition element cation has space for coordination of

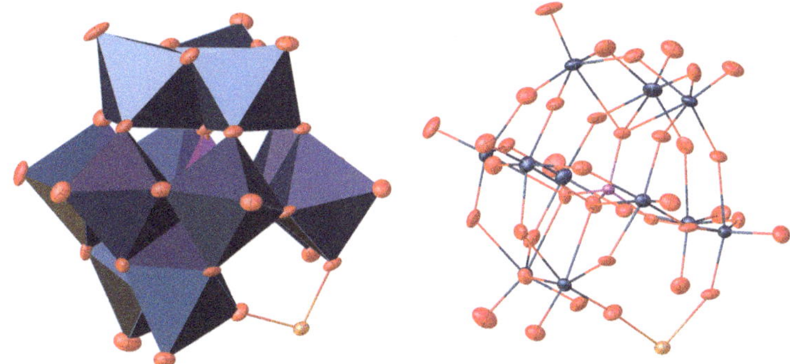

Figure 4.8 Complexed transition element cations in a mono Keggin lacunary-type structure, polyhedron (top) and atomistic (bottom) model. Color code: purple - X (heteroelement) blue - metals, red - O and orange - foreign transition element cation M(II) (*e.g.* Pd(II), Co(II), Cu(II) or Ru(II)). The data were used from the Cambridge Crystallographic Data Centre and Fachinformations-zentrum Karlsruhe Access Structures service database (deposition number: 2056326).

other ligands (*e.g.* H_2O), making those compounds interesting candidates for catalytic applications.[42–44]

4.1.2 The Heteroelement Position

The heteroelement is an essential component for HPA structures. In addition to stabilizing effects, the heteroelement plays an important role in the formation of different HPA structure types. Here, the origin of the heteroelement from the periodic table has a significant role, as the element size of a group differs depending on the period. Elements with small atomic radii from period 3 (such as Si or P) are more likely to form a Keggin-type structure. On the other side, elements with larger atomic radii from period 5 (such as Sb or Te) tend to form Anderson–Evans-type clusters. The heteroelement in a Keggin-type structure is tetrahedrally coordinated, whereas the heteroelement in an Anderson–Evans-type structure is octahedrally coordinated.[34] This is consistent with the trend that octahedral coordination is preferentially formed with elements of larger atomic radii, since larger atoms have more space for the coordination of six ligands.[45] A tetrahedral coordination motif is formed preferentially with elements of smaller atomic radii, as the atom has less space available to coordinate ligands. Elements from period 4 also seem to favor the tetrahedral coordination motif, but the structural tension caused by the larger element inside the cluster leads to the formation of open motifs, such as a Keggin lacunary-type structure. With As as the heteroelement, the formation of a dimeric Keggin lacunary-type structure similar to eqn (4.4) is formed (with M = Na^+), where both monomers are linked *via* alkali cations (see Chapter 4, Section 4.1.1). Elements from period 2 (such as N) or below are too small to participate in POM formation. DFT simulations confirmed that this is an energetic effect. Larger heteroatoms tend to form octahedral coordination geometries because there is more space available around the heteroatom. Smaller heteroelements require significantly more energy to arrange six oxygen atoms around the element in an octahedral configuration, as the oxygen atoms come too close and repel each other. Therefore, it is easier to arrange only four oxygen atoms around the heteroatom. On the other hand, larger heteroelements offer more space, which correlates with a greater affinity for coordinating multiple oxygen atoms. Consequently, significantly more energy would be required to force larger elements to bind only four oxygen atoms. Nitrogen is even so small that it is already energetically demanding to bind four oxygen atoms in a tetrahedral configuration. Octahedral coordination around a nitrogen atom is energetically completely excluded. In the discussion above, it must also be noted that the influences mentioned are also pH dependent. The formation of the Anderson–Evans structure already takes place at pH values of ~5, while Keggin-type structures form at pH values of ~1.[34]

4.1.3 The Ligand Position

The presence of oxo ligands in POMs is a necessary requirement for bonding the metals together and avoiding metal–metal bond formation. However, it is also possible to replace the oxo ligands with other non-metals, whereby the respective structure type remains unchanged. The following ligand exchanged POMs are known:[46]

- Lindqvist-type POMs in which one or two oxo ligands have been replaced by cyclopentadienyl ligands and so by the element carbon.[46–48]
- Functionalization with nitrogen is also possible, for example using nitrido N^{3-} ligands. However, POMs with N^{3-} ligands are theoretical considerations that are still topics of current research.[46]
- A synthetically more accessible option is the imido functionalization R–N^{2-} of POMs. Imido functionalization was investigated for the first time on Lindqvist-type POMs with stoichiometric compositions of $[Mo_6O_{19-x}(NR)_x]^{2-}$ (with $x = 1$ to 6). R is an organic residue, *e.g.* compounds like $R^1{}_3$ P=N–R^2 (with R^1 and R^2 = aryl) a

so-called phosphinimide, R–N═C═O an isocyanate, or an amine R–NH_2 can be considered.[46]

- Hydrazido-functionalization on Lindqvist-type POMs (using hydrazides like H_2N–NR_2, with R═Ar but there are currently no applications known.[46]

The above-mentioned ligand examples are classified as so-called "innocent" ligands (see Figure 4.9). This means that the ligand does not change the oxidation state of the metal. Here, the metal remains in its d^0 configuration, having no available electrons in the d orbitals. A discussion of ligand examples classified as "non-innocent" ligands follows. Here, the ligand is involved in the reduction of the metal, meaning that the ligand is not "innocent".[46]

- Diazenido-ligand (–N=N–R) modification ("non-innocent" ligand behavior) by primary hydrazines. The ligand reduces the metal to its d^4 state, resulting in a M(II) species for Mo(VI) or W(VI), meaning the metal has four electrons available. The bond situation can be understood as a σ donation of the diazonium ligand to the metal (transfer of electron density from the ligand σ orbital to the d orbitals (*e.g.* d_{z^2} or $d_{x^2-y^2}$) of the metal) and π back-donation of the metal to the diazonium ligand (transfer of electron density from the metal d orbitals (*e.g.* d_{xy}, d_{xz} or d_{yz}) to a ligand π orbital). The bond situation is visualized in Figure 4.9.[46]

Such bonding situations are often explained using the donor–acceptor interaction model. In general, electron density can be transferred from a ligand to a metal if σ-MOs of suitable symmetry overlap, *e.g.* the σ-MO of the ligand with a σ-orbital of the metal (d_{z^2} or $d_{x^2-y^2}$). This creates a σ-bond, a process that is called σ-donation. So, the electron density on the metal is increased and the metal can now transfer electron density to the ligand. The metal is in its d^4 configuration and has four electrons already available, two so-called free electron pairs on the metal atom. Here, MOs of suitable symmetry must overlap, *e.g.* the d_{xy}, d_{xz} or d_{yz} orbital of the metal with an antibonding π*-MO of the ligand. This creates a π-bond, resulting in a mesomeric formula with a double bond between the ligand and metal. The process is then called π-backdonation. Both bonds reinforce each other and are therefore synergistic. "Non-innocent" ligands are therefore classified as π acceptor ligands.[40,49,50]

With "innocent" ligands, the metal has no electrons available and is in its d^0 configuration. Here, the σ-MO of the ligand can overlap

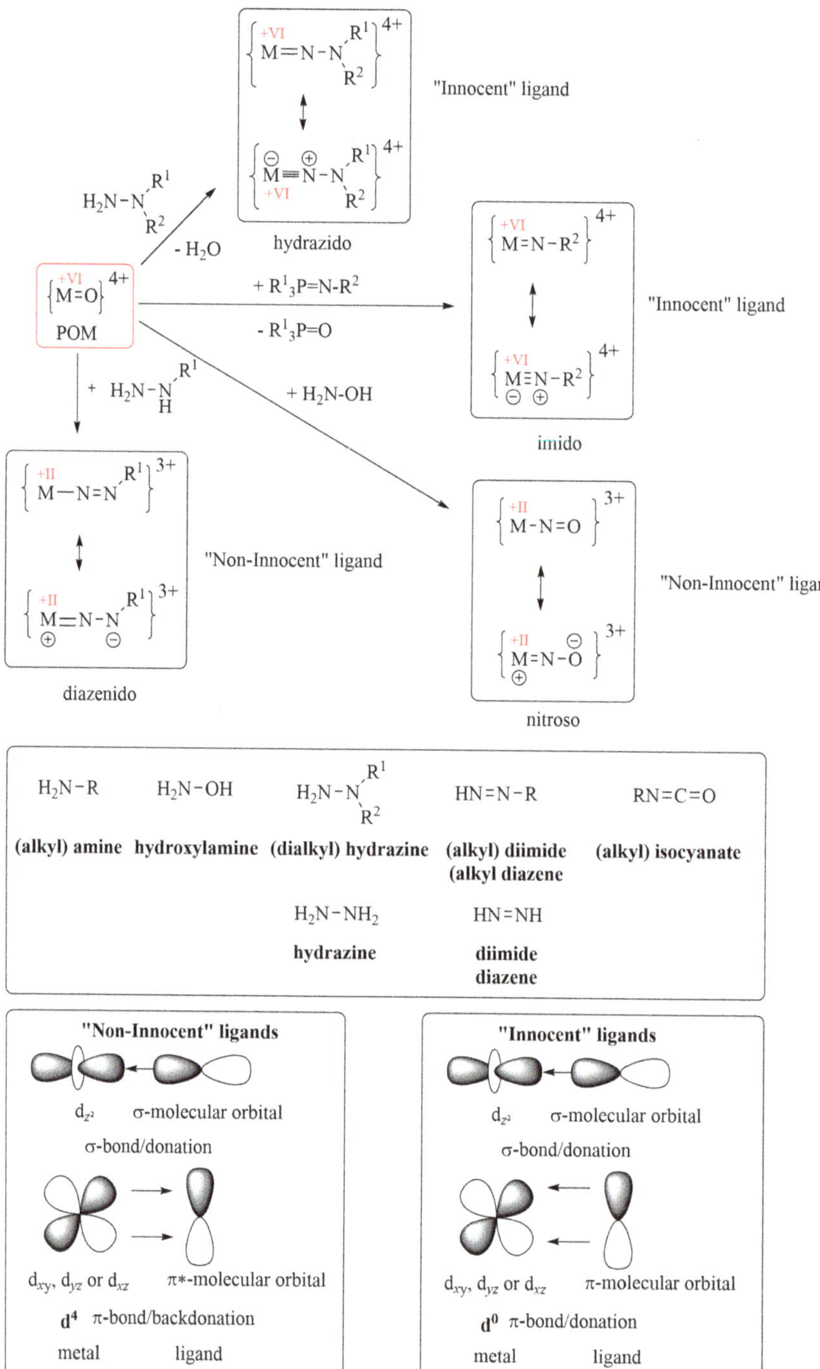

Figure 4.9 Interaction of "innocent" (no RedOx reaction involved) and "non-innocent" ligands (RedOx reaction involved) with the framework metal of a POM.[46]

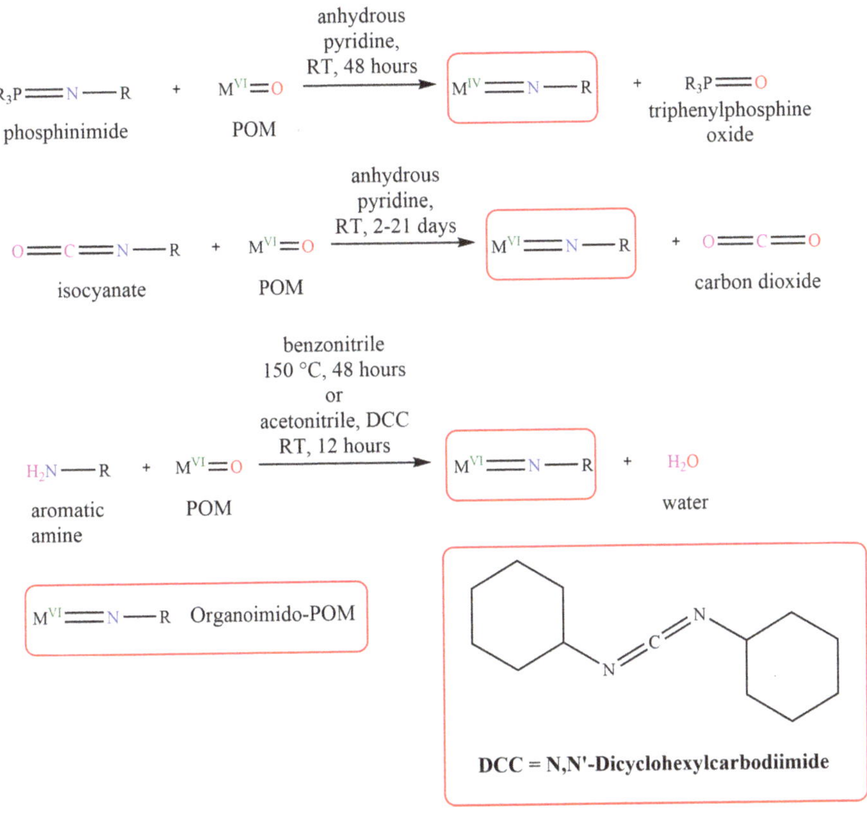

Figure 4.10 Approaches for preparing organoimido functionalized POMs.

with the $d_{x^2-y^2}$ or d_{z^2} orbital of the metal to form a σ-bond (σ-donation). However, the metal cannot transfer electron density back to the ligand, because it has no electrons available in its d^0 configuration. The occupied π-MO of the ligand overlaps with the d_{xy}, d_{xz} or d_{yz} orbitals of the metal and transfers more electron density to the metal, resulting in another π-bond to the metal (π-donation). So, a mesomeric formula can be drawn with a double bond to the metal. However, the second electron pair can be considered as the free electron pair of the nitrogen atom, coordinating to the metal. The "innocent" ligands can be classified here as π-donor ligands.[40,49,50]

Synthetic approaches for organoimido-POMs are shown in Figure 4.10:

- Conversion of phosphinimides with a POM, forming triphenylphosphine oxide (TPPO) as a by-product (the thermodynamic driving force is the formation of TPPO, due to the oxophilicity of phosphorus).[46,51–53]

- Conversion of isocyanates with POMs, forming carbon dioxide (CO_2) as a by-product (the thermodynamic driving force is the formation of CO_2).[46]
- Conversion of aromatic amines with POMs forming water as a by-product. The reaction can be accelerated by activation of the terminal oxo ligand with *N*,*N*′-dicyclohexylcarbodiimide (DCC).[46]

It must be noted that the stability of the POM–imido bond M=N–R is not necessarily protected against hydrolysis. Water can cleave the bond, forming the primary amine R–NH_2 and the POM M=O (restoring the terminal oxo ligand). This is due to the fact that metals have a higher affinity for the element oxygen (oxophilicity) than for other nonmetals. Another ligand exchanged POM is the Anderson–Evans-type compound $[Bi_7I_{24}]^{3-}$, called heptabismuthate.[54] The oxo ligand is replaced by iodine and the framework element is the main-group element Bi. This compound can be prepared, starting from palladium(II) acetate and bismuth(III) iodide in tetrahydrofurane (THF) solution.[54]

4.1.4 Hydration Water

Normally, POMs are prepared in aqueous solution and isolated as multihydrates, according to the general molecular formula $C_x[X_yM_zO_n]\cdot mH_2O$. Typical values for m are 3 to 28[5,6] or, in the case of the molybdenum blue rings, 400.[55] An analytical method to determine the hydration water content is thermogravimetric analysis (TGA). Here, the POM is heated to a defined temperature (*e.g.* 350 °C)[56] and the hydration water evaporates, which is noticeable by a constant loss of mass. The evaporation of water takes place at temperatures just above room temperature (low bonded water) and continues until well above 100 °C, often even above 200 °C (strongly bonded water).[1,56]

The hydration water is incorporated into the crystal structure of the POM and is therefore more difficult to vaporize than free water. Hydration and vaporized water are in an equilibrium, which shifts to the side of the free water if the temperature is constantly increased and/or the pressure is lowered. Evaporation of the water can be prevented with increasing temperature if the TGA experiment is done in a water vapor atmosphere. The equilibrium shifts to the side of the bonded hydration water (Le Chatelier's principle). A typical TGA curve of an experiment, done in a nitrogen (N_2) atmosphere, is shown in Figure 4.11.[1]

The difference between the initial mass (mass of the POM material + hydration water) and the mass constancy (mass of the pure POM material) can be interpreted as the mass of the hydration water. Strongly

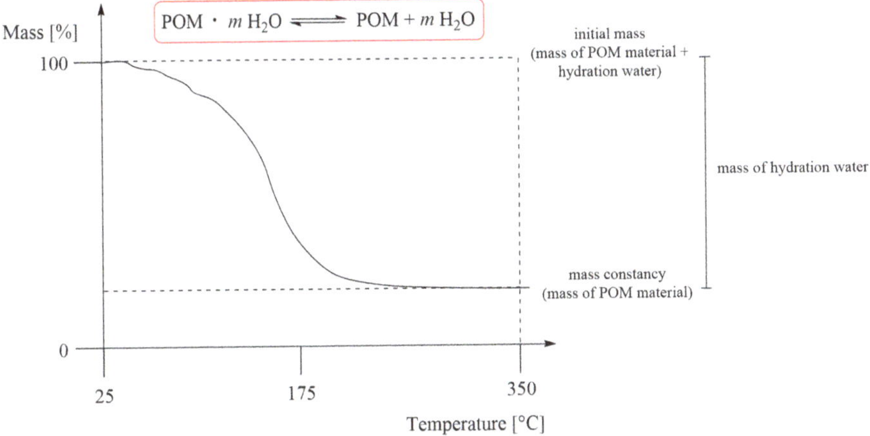

Figure 4.11 TGA curve of a POM. During heating, the hydration water evaporates according to a shift of the chemical equilibrium. The material loses mass until the mass constancy is reached. Initial mass is the mass of the POM material plus hydration water, and mass constancy is the mass of the pure POM material. The difference can be interpreted as the mass of the hydration water.

bonded water molecules are found if they are coordinated to cations (like alkali or transition metal cations). Those water molecules will evaporate only at elevated temperatures >150 °C. In general, the higher the temperature, the more hydration water molecules will evaporate. This effect is driven by an increase in entropy if the water is free.[1,56]

Salts (H^+, $NH_4{}^+$, K^+ and Cs^+ salts) of the POM anions $[PMo_{12}O_{40}]^{3-}$ and $[PVMo_{11}O_{40}]^{4-}$ were investigated using TGA analysis. The following observations were found:[56]

- In the temperature range between 20–170 °C all hydration water molecules were evaporated.
- A slight mass loss was observed in the temperature range between 170–420 °C for $[PMo_{12}O_{40}]^{3-}$ salts and 170–360 °C for $[PVMo_{11}O_{40}]^{4-}$ salts, explained by the release of molecular O_2, which is accompanied by a reduction of V(V) to V(IV) and Mo(VI) to Mo(V).
- For H^+ and $NH_4{}^+$ salts of the anion $[PVMo_{11}O_{40}]^{4-}$, an additional water release (mass decrease) was observed at temperatures of 360 °C with decomposition of the Keggin-type structure. For the same salts of the anion $[PMo_{12}O_{40}]^{3-}$, the decomposition was observed at temperatures around 420 °C.
- A reoxidation step (V(IV) to V(V) and Mo(V) to Mo(VI)) was observed for both salts at around 458 °C (mass increase).

- Complete destruction of the Keggin-type anion and the formation of the oxides P_2O_5, MoO_3 and V_2O_5 was observed at temperatures over 470 °C.

The release of water can be delayed by choosing cations like Cs^+. NH_4^+ salts decompose to oxides at temperatures of about 470 °C, due to the fast step of N_2 formation. The protons form water with the oxo ligands of the POM anion. A decomposition of the NH_4^+ anion is thermodynamically preferred due to the increase in entropy. In general, POM salts are more thermally stable compared to the POM acids, because there are less (for NH_4^+ salts) or no (for K^+ or Cs^+ salts) protons available to form water molecules with the oxo ligands of the cluster (accompanied by full destruction of the structure type). For the acids, a decomposition of the Keggin-type structure and the formation of water was observed at temperatures of 360 °C to 370 °C.[56]

In summary, POMs show a very high stability against temperature, which can be controlled by choosing the correct cation. In this property, POMs are superior to many organic, inorganic and organometallic compounds.

4.2 Crystallographic Properties

Through a detailed crystallographic investigation of POMs, typical values of bond lengths that can be assigned to different groups in a POM can be identified. The assignment of the different groups of bond lengths is demonstrated using the example of a Keggin lacunary-type structure in Figure 4.12.

The idea of the sum of covalent radii assumes that an ideal covalent bond length simply corresponds to the sums of both atomic radii. This means that drastically shortened lengths between two elements can be interpreted as strong covalent bonds (*e.g.* a double bond between two elements). Extended lengths indicate weak interactions, which can be explained by coordinative or ionic interactions.[57]

If the values of different bond lengths for different POM structures are evaluated, the typical bond lengths listed in Table 4.1 can be found.

The shortened bond length of the X–O bonds indicates a strong, covalent bond character, explained by the high oxophilicity of the non-metal heteroelements. Significantly longer bond lengths are found for the M–O_X bonds, indicating a weaker covalent bond. This observation is consistent with the host–guest hypothesis and can be explained by a coordinative interaction between the heteroelement

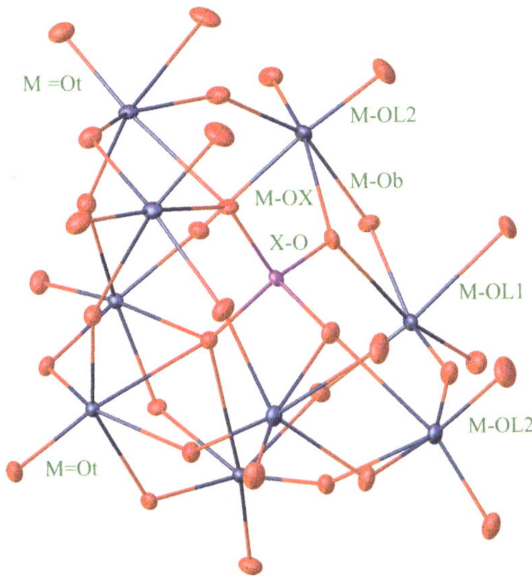

Figure 4.12 Different bond types found in POM-type structures. Color code: purple – X, blue – metals M, and red – O. The data were used from the Cambridge Crystallographic Data Centre and Fachinformationszentrum Karlsruhe Access Structures service database (deposition number: 2205006).

polyhedron and the metalate framework. The length of the M–O_b bond is comparable to the sum of covalent radii, so a covalent bond character can be assumed. A slight shortening results, which is explained by the oxophilicity of the metals, similar to the heteroelements. All terminal M=O_t bonds are significantly shortened, which indicates a stronger covalent character, explained by a double bond. This trend is also found for the lacunary-type oxo ligands M–O_L. However, an alternating trend is found here, with an extended M–O_{L1} and a shortened M–O_{L2} bond motif.[1,2,5,6,34]

For the POM anion–cation interaction, crystallographic observations for POM alkali salts show that the cations are arranged in edge-linked octahedra, with some alkali cations coordinating to the terminal oxo ligands of the POM anion. Furthermore, the alkali cations are partially coordinated by up to six hydration water molecules, with two alkali cations connected by two common oxygen atoms of the hydration water molecules. The situation is shown in Figure 4.13 top, left and right.[6]

Note: hydrogen atoms of the hydration water molecules are not shown. Water molecules that are directly coordinated to cations are responsible for the high temperatures required in a TGA experiment to evaporate the hydration water.[6]

Table 4.1 Typical bond lengths found in POM clusters can be categorized as follows: X–O (heteroelement–oxygen bond), M–O_X (metal–oxygen bond, oxygen is here connected to the heteroelement X), M–O_b (metal–oxygen bond of the metal–metal bridging oxygen atoms), M=O_t (metal–oxygen bond of the terminal oxygen atoms) and two different types of metal–oxygen bonds, only found in lacunary-type structures. The values are compared to the sum of covalent radii. 1 Å = 100 pm = 0.1 nm.[57]

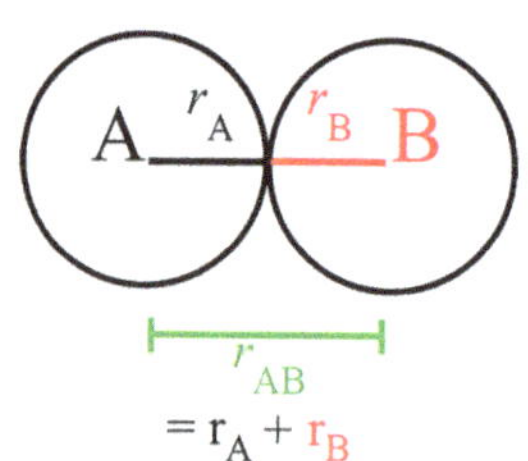

Bond type	Bond length [Å]	Sum of covalent radii [Å][57]
X–O (X = Si, P)	1.5	1.74 (X = P)–1.79 (X = Si)
X–O (X = Sb, Te)	1.9–2.0	1.99 (X = Te)–2.03 (X = Sb)
M–O_X	2.3	2.00 (M = W)–2.01 (M = Mo)
M–O_b	1.9	2.00–2.01
M=O_t	1.7	2.00–2.01
M–O_{L1}	2.2	2.00–2.01
M–O_{L2}	1.7	2.00–2.01

Typical bond lengths of the cation–POM interactions are shown in Table 4.2. The cation–oxo ligand bond length exceeds the sum of covalent radii, which suggests an ionic interaction. This is also in agreement with the bond lengths between the hydration water oxygen atoms and the cations. However, the bond here is more a coordinative type. A similar trend is also found for transition metal cations (here using Co^{2+} as an example), which still exceeds the sum of covalent radii. In POM clusters of the types shown in Figure 4.5, which are isolated as organic ammonium salts and where the Co^{2+} cations additionally coordinate to a free electron pair of a nitrogen atom from an amine, significantly elongated bond lengths are found. Here, a coordinative bond character can be assumed, see Figure 4.13 bottom, left and Table 4.2.[6,32,34]

As shown in Figure 4.13 (bottom, right), the Krebs-type cluster (anion $[Co_2(H_2O)_6\{W(OH)_2\}_2(BiW_9O_{33})_2]^{6-}$) is shown with its different type of bonds. The bond lengths between the M(II) and free oxo ligands are about 2.096 Å, indicating that those oxo ligands can only belong to hydration water molecules. As discussed in Chapter 4, Section 4.1.1, a $[WO_2]^{2+}$ fragment is coordinated to the lacunary-type

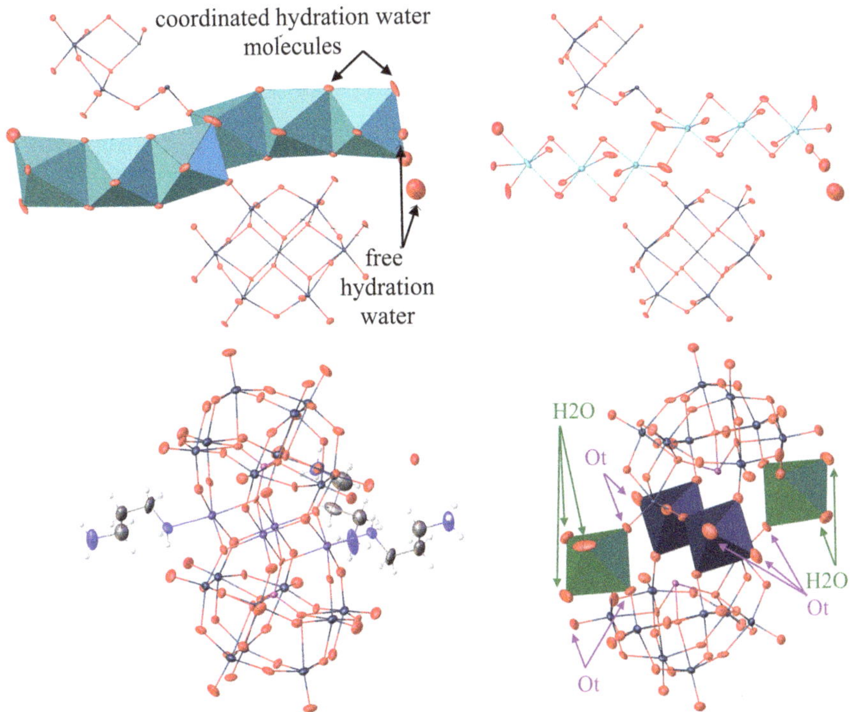

Figure 4.13 (Alkali) cation POM interaction. Polyhedron (top left) and atomistic representation (top right and bottom left). Cation POM interaction in Krebs-type structures (bottom right). Color code: turquoise - alkali cations, green - M(II) (*e.g.* Co(II)/Mn(II)), gray - C, white - H, sky blue - N, purple - X, blue - metals Mo/W, and red - O. The data were used from the Cambridge Crystallographic Data Centre and Fachinformationszentrum Karlsruhe Access Structures service database (deposition numbers: 2293848, 871204 and 730764). Protons of the hydration water molecules are not shown here.

Table 4.2 Typical bond lengths found between cation and POM clusters. The values are compared to the sum of covalent radii. 1 Å = 100 pm = 0.1 nm.[57]

Bond type	Bond length [Å]	Sum of covalent radii [Å][57]
O_t–Na	2.3–2.9	2.18
O_L–Na	2.9	2.18
Na–$O_{hydrate}$	2.4–2.6	2.18
O_X–Co	2.3	1.74
O_L–Co	2.0	1.74
Co–NH_2R	2.2	1.82
M(II)–$O_{hydrate}$	2.1	1.74 (Co)–1.82 (Mn)
$[WO_2]^{2+}$/$[W(OH)_2]^{4+}$	1.9	2.00

cluster (and not a $[W(OH_2)_2]^{6+}$ fragment). This can be seen by investigating the bond lengths between W(VI) and the free oxo ligands of 1.871 Å. It is clear that this bond length is more in the range of a typical $W{=}O_t$ and not in the $W–O_{hydrate}$ bond length range. However, the bond length is elongated compared to typical $W{=}O_t$ lengths of 1.7 Å, indicating that the terminal oxo ligand could be protonated, a so-called hydroxido ligand, forming a $[W(OH)_2]^{4+}$ fragment.[41]

To conclude the crystallographic discussion of POMs, one might ask how large such structures actually are. A typical Keggin-type structure measures 1.03 nm, while a Wells–Dawson-type structure is 1.34 nm tall and 1.03 nm wide. Anderson–Evans-type structures extend to 0.844 nm and are planar with a height of 0.275 nm. A Lindqvist-type structure is 0.832 nm in length. The {Mo154} wheels even reach dimensions of up to 3.6 nm. POMs are nanostructured compounds with considerable spatial dimensions (see crystallographic data sets 2177881, 2216946, 2293848, 2220347 and 745074 in the Cambridge Crystallographic Data Centre and Fachinformationszentrum Karlsruhe Access Structures service database).

4.3 More Solid-state Properties

4.3.1 Structural Isomers

There are different structural isomers of the known POM structure types, which are classified using Greek letters. In general, all isomers are formed in different ratios in a synthetic procedure, whereby the ratios reflect the stabilities of the individual isomers.

4.3.1.1 Keggin-type Structure

The most stable isomer of the Keggin-type structure is the α-structure. Two MO_6 octahedra can be linked together *via* common corners and edges. The edge linkage is the energetically less stable variant due to the higher Coulomb repulsion of the metals. Depending on which edges of the octahedrons are linked, the metal–metal–metal (M–M–M) angle can be 60°, 90°, 120° or 180°, with the 60° angle being the energetically preferred constitution. This isomer is therefore called the α-isomer of the Keggin-type structure. The less stable β-isomer is obtained if a M_3O_{13} triad is rotated by 60°. However, the formation of a Keggin-type structure leads to the formation of both constitutions, so that both isomers are present as a mixture in different ratios.[3,58–61]

4.3.1.2 Wells–Dawson-type Structure

A Wells–Dawson-type structure formally consists of two M_3O_{13} triads (designated M_3) as caps (on top and bottom of the cluster) and two belts consisting of six metals (designated M_6). To create a Wells–Dawson-type structure, two M_6 belts are placed on a M_3 cap followed by another M_3 cap, resulting in the sequence M_3–M_6–M_6–M_3. By rotating the caps and belts, six isomers can be formed, designated α, β, γ, α*, β* and γ*. The order of stability is $\alpha > \beta > \gamma > \gamma^* > \beta^* > \alpha^*$. Formally, a Wells–Dawson-type structure $[X_2M_{18}O_{62}]^{6-}$ is a dimer of two Keggin lacunary-type structure units $[XM_9O_{34}]^{9-}$, designated XM_9. From the typical α-structure, the β-structure is formed by a rotation of the M_3 belt by 60° and the γ-structure by an additional 60° rotation of the opposite M_3 belt. The corresponding α*-, β*- and γ*-structures are formed from the α-, β- and γ-structures by 60° rotations of the opposite XM_9 groups.[62,63]

4.3.1.3 Anderson–Evans-type Structure

While the best-known isomer of the Anderson–Evans-type structure is planar and referred to as the α-isomer, the β-isomer adopts a butterfly-shaped arrangement. The first β-isomer of the Anderson–Evans-type structure was isolated in 1988, with the stoichiometric composition $[H_2SbMo_6O_{24}]^{5-}$.[64] Further representatives have been isolated, including the triol-functionalized derivatives $[Cr\{RC(CH_2O)_3\}_2Mo_6O_{18}]^{3-}$.[65]

4.3.2 Positional Isomers

Positional isomers are only found for TMSPOMs and are a phenomenon that is relevant for TMSPOMs with substitution degrees $x > 1$. For a mono substituted TMSPOM of the Keggin-, Anderson–Evans- or Lindqvist-type structure, it makes no difference which MO_6 octahedron contains the foreign element. All positions are therefore the same (equivalent), which means that there is only one isomer for this TMSPOM species. In an already mono substituted TMSPOM, the remaining positions are no longer equivalent. A further foreign element can now occupy the position directly neighboring to the first foreign element or a position significantly further away. The resulting isomers are referred to as positional isomers. Using the example of a Keggin-type structure, there is only one isomer for $x = 0$ or 1, and a total of 5 for $x = 2$, 13 for $x = 3$, 27 for $x = 4$, 38 for $x = 5$ and 48 for $x = 6$. All five positional isomers for a Keggin-type structure with $x = 2$ are shown in Figure 4.14.[66,67]

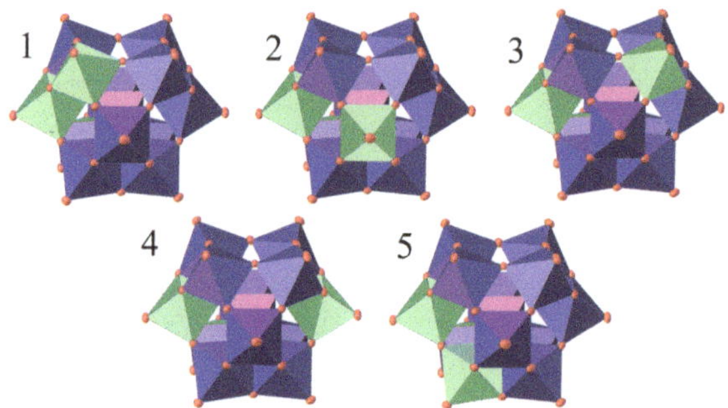

Figure 4.14 Positional isomers for a Keggin-type structure with a substitution degree of $x = 2$. Color code: purple – heteroelement, blue – metals, green – foreign element and red – oxygen.[66,67]

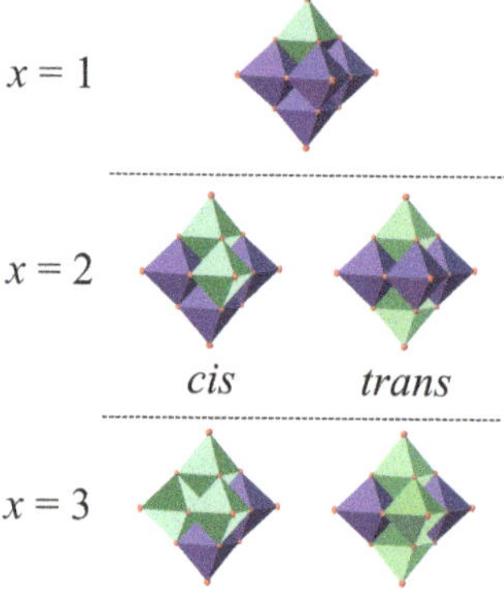

Figure 4.15 Positional isomers for a Lindqvist-type structure with a substitution degree of $x = 1$ to 3 Color code: blue – metals, green – foreign element and red – oxygen.

In isomers 1 and 5, both foreign elements are neighboring. In isomer 1, both foreign element octahedrons are edge sharing, whereas in isomer 5 they are corner sharing.[66,67]

For a Lindqvist-type TMSPOM, there are only two possible isomers when $x = 2$. In one isomer, both foreign elements are neighboring (the *cis*-isomer) and in the other isomer the foreign elements are as far away as possible (the *trans*-isomer), as shown in Figure 4.15.[68–70] For $x = 0$ and 1, only one isomer is found and or $x = 3$, two isomers are possible in which all foreign elements are adjacent to each other.[68–70]

There is no selectivity during the synthetic procedure to control which isomers are formed. The distribution of the isomers depends on the probability (for $x = 2$, the formation of a certain isomer is 1/5 for a Keggin-type and 1/2 for a Lindqvist-type) and on the stability of the respective isomer. If an isomer is not stable, it will not be formed.[66,67]

In the Keggin-type structure (Figure 4.14), the distance between two directly neighboring foreign elements is 3.35 Å (edge-sharing octahedra, **1**) or 3.76 Å (corner-sharing octahedra, **5**). In structure **2**, the distance is 5.01 Å; in structure **3**, it is 6.09 Å; and in structure **4**, it even reaches 7.07 Å. Structure **4** is the isomer in which both foreign elements are as far apart as possible (see Cambridge Crystallographic Data Centre and Fachinformationszentrum Karlsruhe Access Structures service database, deposition number: 2177881). For the Lindqvist-type structure (Figure 4.15), the approximate distance between the foreign elements is 3.26 Å in the *cis* isomer and 4.64 Å in the *trans* isomer (see Cambridge Crystallographic Data Centre and Fachinformationszentrum Karlsruhe Access Structures service database, deposition number: 2220347).

4.4 Chemical Bonds in Polyoxometalates

The chemical bond situations found in POMs can be derived from the trends discussed in Chapter 1; Chapter 2, Section 2.3; and Chapter 4, Section 4.1.3 (see Figure 4.16).

- A σ-atomic orbital ($2p_z$) of the terminal oxo ligand forms a σ bond with the d_{z^2} orbital of the d^0 framework element and transfers electron density to the metal (σ donation). An antibonding π atomic orbital ($2p_x$ or $2p_y$) of the oxo ligand can now overlap with the d_{xy} orbital of the metal and transfer electron density to the metal *via* a π bond (π donation). A M=O double bond is formed. This results in a shortened M=O bond length found in crystallographic data (see Chapter 4, Section 4.2).[49,50]
- The bridging oxo ligand forms a σ bond with its σ-MO and the $d_{x^2-y^2}$ orbital of the metal and transfers electron density to the metal. A M–OM single bond is formed. π MOs of the oxo ligands compete with the d_{xy} orbital. If the bridging oxo ligand transfers electron density to the d_{xy} orbital of the metal *via* its π-MO, a M=OM double bond is formed. This reduces the bond order (BO) of the terminal oxo ligand to one, which no longer overlaps with the d_{xy} orbital. The result is a M–O single bond. However, this mesomeric structure is of no significance (see the shortened bond lengths for terminal oxo ligands, Chapter 4, Section 4.2).[49,50]
- The oxo ligand of the heteroelement polyhedron competes with the terminal oxo ligand for the d_{z^2} orbital of the metal. Since the terminal oxo ligand shows a greater *trans* influence, only σ bonds

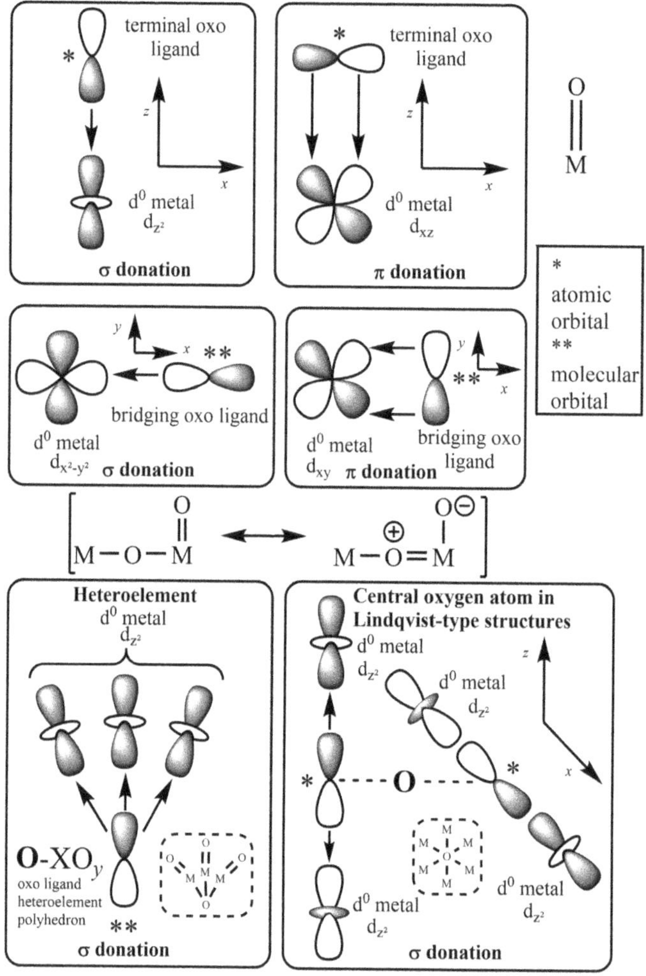

Figure 4.16 Explaining the chemical bond situation in POMs. Oxo ligands are π donor ligands, meaning that they can form a σ bond (σ donation) and π bond (π donation).

are relevant for bond formation, associated with extremely extended bond lengths (see Chapter 4, Section 4.2). In Keggin- or Wells–Dawson-type structures (also in the Waugh-type structure), a σ-MO of the heteroelement oxo ligand competes with three d_{z^2} orbitals of three metals. The oxo ligand of the heteroelement polyhedron formally coordinates to three metals. BOs can be interpreted as 1/3. This is in agreement with the extremely extended bond lengths found for XO–M. On the other side each oxo ligand of the heteroelement in the Anderson–Evans- or Strandberg-type cluster competes for the d_{z^2} orbital of two framework metals (BO is 1/2). In Lindqvist-type POMs, three sigma MOs of the central oxo ligand (p_x, p_y and p_z) compete with six metal d_{z^2} orbitals. Here, the BOs can be interpreted as 1/3. In sum, the oxo ligand forms two covalent σ bonds with six metals (6 × 1/3). This concept is in agreement with the significantly elongated bond lengths found in crystallographic data for Lindqvist-type POMs, especially between the central oxo ligand and the metals.[49,50]

- In Anderson–Evans-type structures, there are two terminal oxo ligands that form double bonds with the metal center. The second oxo ligands binds *via* the $d_{x^2-y^2}$ orbital, for example along the *x*-direction. The σ-MO of the oxo ligand overlaps with the $d_{x^2-y^2}$ orbital of the metal to form a σ bond (σ donation). Additionally, the π MO of the oxo ligand overlaps with the d_{yz} orbital (π donation), contributing to the formation of the double bond.[49,50]

4.5 Polyoxometalates in Solution

Various entropy-induced dissociation equilibria of POMs have been described, in particular for the TMSPOMs in aqueous media (see eqn (4.14) and eqn (4.15)):[3,4,71]

$$2[PV_xM_{12-x}O_{40}]^{(3+x)-} \rightleftharpoons [PV_{x-1}M_{12-x+1}O_{40}]^{(3+x-1)-} + [PV_{x+1}Mo_{12-x-1}O_{40}]^{(3+x+1)-} \quad (4.14)$$

$$[PV_xM_{12-x}O_{40}]^{(3+x)-} \rightleftharpoons nVO_2^{+} + [PV_{x-n}M_{12-x}O_{40-2n}]^{(3+x+n)-} \quad (4.15)$$

The dissociation equilibria show that two molecules of a foreign element-substituted POM can dissociate to one molecule of the higher and one molecule of the lower substituted species (eqn (4.14)). In the next step, the higher and lower substituted species can dissociate again, following this equilibrium. Another dissociation pathway is the release of the foreign element cation, forming a kind of lacunary-type species as a by-product (eqn (4.15)). For higher charged foreign metal cations, oxo compounds of the foreign element are dissociated, *e.g.* for V(V) the VO_2^+ cation. Note: for the reduced V(IV) containing species, VO^{2+} cations are released. In aqueous solution, the equilibria shift to the side of the dissociation products, due to the increase in entropy. After or during evaporation of the water, the equilibrium shifts back to the intact POM species. So, the equilibria are always reversible.[3,4,71]

Water contains one oxygen atom. Driven by the oxophilicity of the metals, water can react with the metal atom and split a M–O–M bond according to eqn (4.16):

$$\text{M–O–M} \overset{+H_2O}{\rightleftharpoons} \text{M–O}^{\ominus} + \text{M–O}^{\ominus} + 2\text{H}^+ \tag{4.16}$$

For this dissociation, an associative or dissociative mechanism can be assumed.

Dissociative mechanism eqn (4.17):

$$\text{M–O–M} \rightleftharpoons \text{M–O}^{\ominus}\text{M}^{\oplus} \overset{+H_2O}{\rightleftharpoons} \text{M–O}^{\ominus} + \text{M–O}^{\ominus} + 2\text{H}^+ \tag{4.17}$$

- In the first step, the M–O–M bond is split (rate-determining step).
- In the second step, water is associated.

Associative mechanism eqn (4.18):

$$\text{M–O–M} \overset{+H_2O}{\rightleftharpoons} \text{M–O–M}^{\ominus}\text{–}\overset{\oplus}{\text{O}}\text{H}_2 \rightleftharpoons \text{M–O}^{\ominus} + \text{M–O}^{\ominus} + 2\text{H}^+ \tag{4.18}$$

- In the first step, a water ligand is associated (rate-determining step).
- In the second step, the M–O–M bond is split.

For the dissociation of VO_2^+ cations from V(v) substituted clusters, a dissociative mechanism with the involvement of water can be assumed, as shown in Figure 4.17.[72]

- In the first step, the V–OM bonds are split to form coordination sites for a water ligand.
- In the second step, water is coordinated by the V(v) center to form a VO_2^+ cation and free protons.

The cleavage of the V–O_X bond to the heteroelement should take place in the last step (blue arrow) and is driven by the *trans* influence in the ground state.[72]

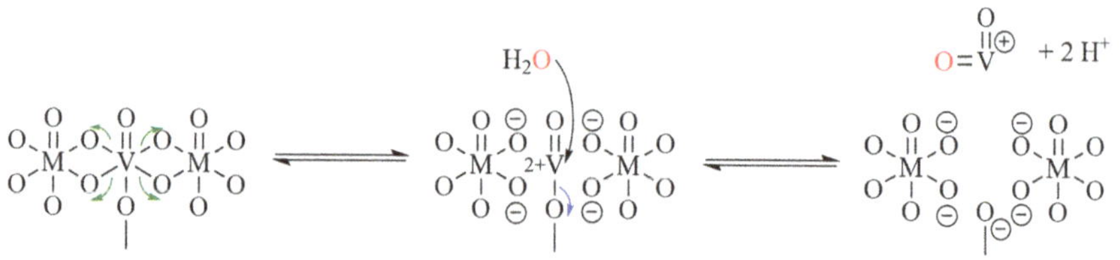

Figure 4.17 Dissociation mechanism of V(v) substituted clusters. (1) V–OM bonds are split/dissociated, (2) a water ligand is added/associated on the V(v) center to form a VO_2^+ cation.

In general, dissociation equilibria depend on:

- the pH value (equilibria can be shifted by adjusting the acidity);
- the temperature; and
- the solvent (in water a high degree of dissociation is observed, in organic solvents the dissociation equilibria are suppressed).

4.6 Stability of Polyoxometalates

In the previous chapters, the following aspects of POM stability were discussed:

- Thermal stability. POMs are stable against increasing temperature. Depending on the cation, decomposition to the element oxides takes place at temperatures >300 °C. Using alkali cations, the decomposition can be delayed. This effect is most pronounced for Cs^+ cations. With H^+ or NH_4^+ cations, the decomposition takes place at lower temperatures.[56]
- Stability is given by a heteroelement, because the heteroelement acts as a template.[34]
- Stability is given by the structure type (different structure types have different stabilities).[6,34,73]
- Stability of POMs is given by the ligands, especially for oxo ligands due to the high oxophilicity of the metals. If oxo ligands are replaced by other non-metals, the M–X–M bond strength decreases due to lower binding affinity to these elements. This makes the M–X–M bonds more sensitive to hydrolysis by water molecules, as the binding affinity of the metals to oxygen is higher.[46]
- The strongest factor is the pH value. POMs are more stable in acidic media. With increasing pH, the M–O–M bonds are split and the POMs decompose to MO_4^{2-} anions. The only anions that are not stable in acidic pH media are $[M_6O_{19}]^{8-}$ with M = Nb or Ta. Those anions are prepared in basic media. Decomposition takes place in acidic pH media.[5,10]
- The charge of the POM anions. For foreign element-substituted POMs, where the framework elements are substituted with foreign elements of lower oxidation states, the charge increases with increasing degree of substitution. The higher the anionic charge, the more negative charge has to be distributed across the POM cluster while maintaining an approximately constant size of the cluster. This reduces its stability, and decomposition

takes place more easily. Therefore, degrees of substitution cannot have any arbitrary values. However, this should not be confused with the limitation of the degree of substitution resulting from the vacancies of a lacunary-type structure.[1,2,4,6]

4.7 Spectroscopic Properties

4.7.1 Vibrational Spectroscopic Properties

The observed vibrational modes for POM clusters can be classified for different types of structure as follows:

- For Keggin-type clusters:[74]
  - Simple P–O vibrations along one bond.
  - Symmetric and asymmetric vibrations of the PO_4 tetrahedron.
  - Vibrations of the terminal oxo ligands $M{=}O_t$.
  - M–O–M vibrations of the edge and vertex sharing oxo ligands.
- For Wells–Dawson-type clusters, similar vibrational modes compared to Keggin-type species are observed.[75]
- For Anderson–Evans-type clusters:[6,34,76]
  - Simple symmetric and asymmetric Te/Sb–O vibrations along one bond.
  - Symmetric and asymmetric vibrations of the Te/SbO_6 octahedron.
  - Vibrations of the terminal oxo ligands $M{=}O_t$.
  - M–O–M vibrations of the edge sharing oxo ligands.
- For Lindqvist-type clusters:[77,78]
  - Vibrations of the terminal oxo ligands $M{=}O_t$.
  - Symmetric and asymmetric vibrations of the M–O–M units.

Vibration bands can be detected experimentally using infrared (IR) and Raman spectroscopy. The list of the different vibration bands shown above is a simplified representation. In reality, the situation is much more complicated and can only be captured by theoretical calculations. Characteristic vibrational frequencies for the above-mentioned structure types are listed in Table 4.3.

4.7.2 Nuclear Magnetic Resonance Spectroscopic Properties

The relevant nuclei for a Nuclear Magnetic Resonance (NMR) investigation of POMs are listed in Table 4.4.[80–86]

Table 4.3 Characteristic vibrational frequencies (in cm^{-1}) observed in IR spectroscopy for Keggin-, Wells–Dawson-, Anderson–Evans- and Lindqvist-type structures.

Bonding unit	Keggin-type $[PMo_{12}O_{40}]^{3-}$/ $[PW_{12}O_{40}]^{3-}$	Wells–Dawson-type $[P_2W_{18}O_{62}]^{6-}$	Anderson–Evans-type $[TeMo_6O_{24}]^{6-}$/$[TeW_6O_{24}]^{6-}$	Lindqvist-type $[Nb_6O_{19}]^{8-}$
P–O	1059/1073	1091	—	—
$(PO_4)_{bending,asym}$	590–600		—	—
$(PO_4)_{bending,asym}$	480		—	—
$M{=}O_t$	962/973	963	946, 886	849, 828
$(M–O–M)_{vertex}$	877/904	911	—	—
$(M–O–M)_{edge}$	744/756	778	670–690	—
$(M–O–M)_{asym}$	—	—	—	656
$(M–O–M)_{sym}$	—	—	—	510
$(Te/Sb–O)_{sym}$	—	—	650	—
$(Te/Sb–O)_{asym}$	—	—	550–570	—
$(Te/SbO_6)_{bending,asym}$	—	—	430–470	—
$(Te/SbO_6)_{bending,sym}$	—	—	220–270	—

Table 4.4 NMR properties of some nuclei that can be measured in POM chemistry.[79]

Property	^{17}O	^{31}P	^{125}Te	^{51}V	^{95}Mo	^{183}W
Spin	5/2	1/2	1/2	7/2	5/2	1/2
Natural abundance [%]	0.038	100	7.07	99.75	15.92	14.31
Chemical shift range [ppm]	1160 (−40 to 1120)	430 (−180 to 250)	5800 (−1400 to 3400)	1900 (−1900 to 0)	4300 (−2000 to 2300)	6720 (−4670 to 2050)
Reference compound	D_2O (0 ppm)	85% H_3PO_4 in H_2O (0 ppm)	90% Me_2Te in C_6D_6	90% $VOCl_3$ in C_6D_6	2 M Na_2MoO_4 in D_2O (0 ppm)	1 M Na_2WO_4 in D_2O (0 ppm)
Receptivity relative to ^1H at natural abundance	1.11×10^{-5}	6.63×10^{-3}	1.64×10^{-4}	0.380	5.21×10^{-4}	1.07×10^{-5}
Receptivity relative to ^{13}C at natural abundance	0.0650	37.7	0.961	0.818	3.06	0.0631
T_1 relaxation of reference [s]	0.02	0.50	~2.0	0.00500	0.81	5.0

NMR measurements of POMs can be performed in solid-state or in solution. Normally, a POM NMR measurement in solution takes place in pH adjusted aqueous media or in deuterated acetonitrile for POMs modified with organic cations. For measurements in aqueous media up to 10% of a deuterated solvent can be added, *e.g.* D_2O, acetonitrile-d_3, acetone-d_6 or DMSO-d_6, to ensure the "solvent lock" during measurement.[80–86]

During NMR measurements of the different nuclei in Table 4.4, the natural abundance of the elements must be considered. The lower the value, the longer (more scans) the sample must be scanned. In the case of the isotope ^{17}O, specific enrichment of the isotope is required in order to obtain meaningful spectra. When planning a NMR measurement, the relaxation times must also be considered in order to keep the measuring time as short as possible. The nuclei ^{17}O, ^{51}V possess quadrupole moments (spin > 1/2), which interfere with NMR measurements by causing signal broadening and intensity loss. Nevertheless, the signal integral remains constant.[87]

For foreign element-substituted TMSPOMs, the following trends can be derived from NMR spectra:

1. Positional isomers. Each positional isomer results in an NMR signal. For the unsubstituted and the monosubstituted POMs, only one NMR signal can be observed. In di-, tri-, tetra-, penta- and hexasubstituted Keggin-type POM anions, 5, 13, 27, 38 and 48 positional isomers exist, resulting in 5, 13, 27, 38 and 48 NMR signals.[3,66,67,88]
2. Constitutional isomers. Even if the α-isomer is the thermodynamically most stable isomer, the β-isomer can still be identified in the NMR spectrum. Peaks of low intensity can be attributed to β-isomer peaks.[3,58–61]
3. Dissociation equilibria. The dissociation equilibria also have an effect on the NMR spectra and therefore show various peaks that can be assigned to dissociation fragments of different isomers.[3,71]

The arguments discussed above illustrate that interpreting TMSPOM spectra can become very complicated, especially for V(v) substituted TMSPOMs. While neighboring H or C atoms of organic compounds can be identified by *J*-coupling, this is not common for POMs. Therefore, structure elucidation using NMR spectroscopy alone is not possible. A simplification of the NMR spectra becomes apparent as soon as individual positional isomers are not stable and are therefore not formed, or if dissociation of the POM (*e.g.* in organic solvents) does not take place.

A further complication is found if paramagnetic elements (*e.g.* Co(II) or Fe(III)) are present. This results in a broadening of the baseline or in completely uninterpretable spectra.[1,2,5,89] Electron spin resonance (EPR) spectroscopy is an alternative for those samples.[90]

4.7.3 Photometric Spectroscopic Properties

TMSPOMs can be investigated with ultraviolet-visible (UV-Vis) spectroscopy, due to their intense colors.

In general, color results if the light of a defined wavelength interacts with a compound. Electrons are excited from an energetic state of low energy to a state of high energy, which is partly or not occupied by other electrons. However, the human eye can only see the complementary color of those wavelengths. For transition elements, the transition of an electron from a d-orbital to another d-orbital (d–d transition) is responsible for the color. The energy required for this transition is dependent on the splitting of the octahedral ligand field Δ_{oct}. Considering a d^0 cation, there are no electrons available for a transition, so there is no color in this ion. A d^1 hexaaquatitan(III) complex $[Ti(H_2O)_6]^{3+}$ has a violet color, because Ti(III) has one electron, that can be excited from the t_{2g} (d_{xy}, d_{yz}, d_{xz}) to the e_g orbitals ($d_{x^2-y^2}$, d_{z^2}).[49]

However, sometimes color is observed for d^0 elements, *e.g.* in $KMnO_4$. This illustrates that there is another effect involved – the so-called ligand-to-metal charge transfer (LMCT) or, in general, the charge transfer (CT). CT encompasses LMCT (charge is transferred from a ligand to the metal), ligand–ligand transfer (LLT) and metal-to-ligand charge transfer (MLCT). In organometallic compounds, only the ligand is responsible for the color, resulting from a π–π* transition, a so-called LLT. For a detailed discussion the selection rules must be considered, because they decide the intensity of the resulting color. To excite an electronic transition, light of a defined wavelength needs to be absorbed. The extinction coefficient ε defines how much light of a defined wavelength is absorbed by a 1 mol L^{-1} solution of the analyte (analyte = the compound that is investigated) contained in a cuvette of 1 cm thickness. The higher the value of ε, the more intense the color.[49]

The following selection rules limit the possible charge transfers:

1. Spin prohibition. The multiplicity M is not allowed to change during a transition, with $M = 2S + 1$. M is unchanged if the spin S of an electron does not change. Values of S can only be $+0.5$ and -0.5. Considering the Fe(III) d^5 cation in a high-spin configuration, each electron occupies one of the five d-orbitals (Figure 4.18). All

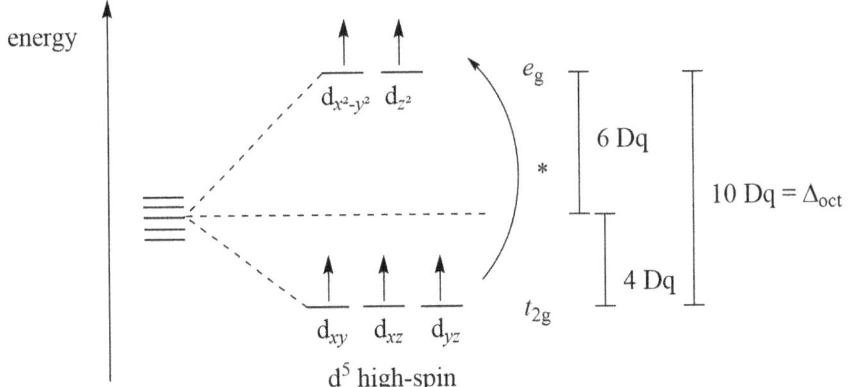

Figure 4.18 In a d^5 high-spin state of an octahedral ligand field no d–d transition can take place with spin retention. If one electron should be excited from the t_{2g} to the e_g orbitals, the electron needs to change its spin *S*. This is forbidden by the spin prohibition. The d–d transition will not take place.[49]

electrons have the same spin (same value of *S*). For a d–d transition, an electron from the t_{2g} orbitals needs to be excited into the e_g orbitals. According to the Pauli principle, the electron needs to change its spin. However, the spin-prohibition does not allow this transition, resulting in a colorless cation.

2. The Laporte prohibition. Here, there are two questions that need to be answered.[49]
    1. After the excitation, is the electron in an orbital with a different parity compared to the orbital from which the electron started? If yes, the excitation is allowed, *e.g.* the transition of an electron from a p- to a d-orbital. If not, the transition is not allowed. The parity of an orbital represents the behavior of the orbital doing a point-reflection. The question, "Has the orbital an inversion center?" needs to be answered. If yes and the orbital does not change after point-reflection, then the parity is "even" (*e.g.* s- and d-orbitals). If not, the parity is "odd" (*e.g.* p- and f-orbitals). A transition is only allowed if an electron is excited from an "even" (p-) orbital to an "odd" (d-) orbital (*e.g.* a p→d transition is allowed, a p→p or a d→d is forbidden). Electronic excitations can be indicated by an arrow "→".[49]
    2. Does the compound have an inversion center? If no, the excitation is allowed (*e.g.* tetrahedral compounds). If yes, the excitation is not allowed (*e.g.* octahedral compounds) (Figure 4.19).[49]

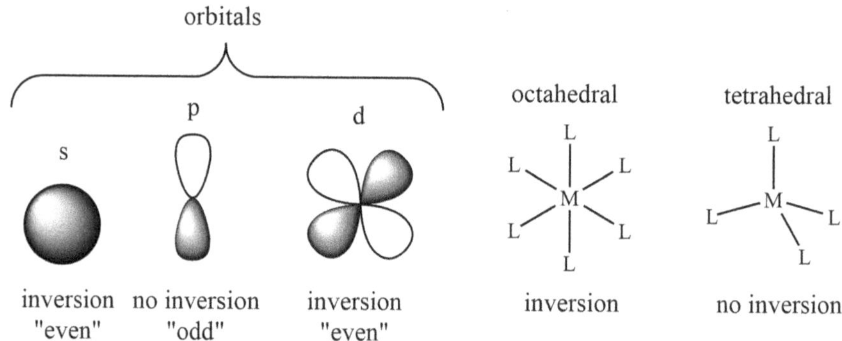

Figure 4.19 "Even" and "odd" character of the orbitals. There is an inversion center for octahedral complexes but not for tetrahedral complexes.

The Laporte prohibition is not as strict as the spin prohibition. This is the reason d–d transitions ("even" to "even" orbitals) are observed. However, electronic transitions in octahedral compounds are known. Those transitions result from vibrations of the octahedra in which the octahedral symmetry is lifted for a short time, making an excitation possible.[49]

In summary, if an electronic transition is spin-prohibited, the excitation will not take place. A transition that is Laporte-prohibited will take place, but the color has a low intensity with a low value of ε. For a Laporte-allowed transition, the excitation is observed, and the color has a very high intensity with a high value of ε. Examples for Laporte-allowed transitions are CTs (*e.g.* LMCTs) with an excitation from an "even" (p-) orbital to an "odd" (d-) orbital. LMCTs are responsible for the intense color of the anion MnO_4^-. The transition of an electron is spin- and parity-allowed (excitation from an "even" (p-) oxo ligand-orbital to an "odd" (d-) Mn-orbital). There is no inversion center, making the color very intense (high value of ε).[49]

LMCTs can be understood as a reduction of the metal by the ligand. However, a LMCT will only be observed if the metal is easy to reduce. The more unstable the oxidation state of the metal, the easier a metal can be reduced. For MnO_4^-, with Mn in its oxidation state of +7, the high oxidation state is unstable and the Mn(VII) species can be easily reduced by the oxo ligands. However, for the anion VO_3^-, the element V is very stable in its oxidation state of +5, meaning that no reduction and thus no LMCT by the oxo ligands will take place. VO_3^- is colorless.[49]

In the UV-Vis spectra of POMs, very intense LMCT bands are found for the framework/foreign elements at wavelengths between 200 and 310 nm. If all metals are present in their highest oxidation states, no d–d transitions will be observed in the range between 450 and 862 nm. If POMs get partly reduced or foreign elements in low oxidation states are present, d–d transitions can be observed with a very low intensity.[89,91–95]

Table 4.5 Absorption bands [nm] in the POM anions $[PMo_{12}O_{40}]^{3-}$ and $[PV_2Mo_{10}O_{40}]^{5-}$ found in the UV-Vis data.[3,74]

Band	$[PMo1_2O_{40}]^{3-}$	$[PV_2Mo_{10}O_{40}]^{5-}$
LMCT O → Mo(VI)	214	215
LMCTO → V(V)d–d	—	308
	450, 540	488, 565, 590
V(IV)–O–Mo(VI)	—	649, 667, 610
V(V)–O–Mo(V)d–d	—	658, 750
	800, 862	—
Mo(V)–O–Mo(VI)	1160	—

In addition to d–d and LMCT bands, so-called intervalence charge transfer (IVCT) bands are observed, in which electrons are transferred from one metal to a neighboring metal, which are connected to each other by oxo ligands.[74] The absorptions found for Keggin-type anions $[PMo_{12}O_{40}]^{3-}$ and $[PV_2Mo_{10}O_{40}]^{5-}$ are summarized in Table 4.5.[3,74]

In summary, UV-Vis spectra of POMs can be explained as follows:

- In a fully oxidized POM, all metals are present in a d^0 configuration (highest oxidation states), meaning that no d–d transition can be observed. The only observed bands are LMCTs, transfers of electrons from the oxo ligands to the metals (formal reduction). Those bands are very intense due to the Laporte prohibition (transfer of an electron from an "odd" p to an "even" d orbital) and the spin prohibition (spin of the excited electron is unchanged).
- In a reduced POM, one or more electrons are transferred into the (empty) d orbitals, resulting in a d^1/d^n ($n>1$) configuration of the metal. In addition to the LMCTs, d–d transitions can now take place. Those transitions have a low intensity due to the Laporte prohibition (transfer of an electron from an "even" d to another "even" d orbital). The spin prohibition (spin of the excited electron is unchanged) is met.

4.8 Acidity of Polyoxometalate Acids

The best-known POM acids are $H_3[PMo_{12}O_{40}]$ and $H_3[PW_{12}O_{40}]$, whose protons can stepwise dissociate in an aqueous medium (see eqn (4.19)):

$$\begin{aligned} H_3[PM_{12}O_{40}] + 3H_2O &\overset{k_1}{\rightleftharpoons} H_2[PM_{12}O_{40}]^- + H_3O^+ + 2H_2O \\ &\overset{k_2}{\rightleftharpoons} H[PM_{12}O_{40}]^{2-} + 2H_3O^+ + H_2O \\ &\overset{k_3}{\rightleftharpoons} [PM_{12}O_{40}]^{3-} + 3H_3O^+ \end{aligned} \tag{4.19}$$

Table 4.6 pK_a values of $H_3[PMo_{12}O_{40}]$ and $H_3[PW_{12}O_{40}]$ in aqueous and acetonitrile solution according to Reddy and Dias *et al.*[96,97]

POM	p$K_{a,1}$	p$K_{a,2}$	p$K_{a,3}$
$H_3[PMo_{12}O_{40}]$ in water	2.40	4.31	5.46
$H_3[PW_{12}O_{40}]$ in water	1.60	3.54	7.16
$H_3[PMo_{12}O_{40}]$ in acetonitrile	−2.22	0.65	2.37
$H_3[PW_{12}O_{40}]$ in acetonitrile	−6.07 (−5.57[a])	−3.62 (−3.47[a])	−1.24 (−3.40[a])

[a]Values according to Dias *et al.*[97] All other values according to Mohan Reddy *et al.*[96]

The more protons dissociate, the stronger the anionic charge of the remaining POM anion and the stronger the basicity of the anion. Therefore, the acidity of the first proton is the strongest and the acidity decreases significantly for the second and third protons.

Table 4.6 summarizes the measured pK_a values of $H_3[PMo_{12}O_{40}]$ and $H_3[PW_{12}O_{40}]$ in aqueous and acetonitrile solution according to Reddy *et al.* and Dias *et al.*[96,97] The first and the second pK_a value of $H_3[PW_{12}O_{40}]$ are lower than for $H_3[PMo_{12}O_{40}]$, meaning that $H_3[PW_{12}O_{40}]$ is a stronger acid. The third pK_a value of $H_3[PMo_{12}O_{40}]$ is lower than for $H_3[PW_{12}O_{40}]$, indicating that the third proton of $H_3[PMo_{12}O_{40}]$ is more acidic. This observation can be explained by the different nature of both framework elements. W is a much bigger atom than Mo, meaning that the negative charge of the anion can be distributed over a bigger cluster, stabilizing more negative charge.[96]

It has been shown that in the solid-state structure of POM acids, the protons are more likely to be located on the bridging oxo ligands rather than on the terminal oxo ligands. For the unsubstituted POMs listed in Table 4.6, larger distances between protons within the POM framework are generally favored. In contrast, for the substituted anions of the type $[PV_xM_{12-x}O_{40}]^{(3+x)}$, the protons tend to localize at the nucleophilic sites adjacent to the vanadium atoms. Up to three protons can be coordinated around a single vanadium atom.[98]

Abbreviations

POM	Polyoxometalate
RedOx	Reduction/oxidation
HPA	Heterpolyanion/heteropolyacid
HSAB	Hard and soft acids and bases
TBA	Tetrabutylammonium
IL	Ionic liquid
pH	Negative decadic logarithm of the hydrogen ion concentration

DFT	Density functional theory
MO	Molecular orbital
TPPO	Triphenylphosphine oxide
DCC	*N*,*N*′-Dicyclohexylcarbodiimide
THF	Tetrahydrofurane
TGA	Thermogravimetric analysis
Å	Ångström (1 Å = 10^{-10} m)
TMSPOM	Transition metal-substituted polyoxometalate
BO	Bond order
IR	Infrared
NMR	Nuclear magnetic resonance
DMSO	Dimethyl sulfoxide
d	Deuterated
ESR	Electron spin resonance
UV-Vis	Ultraviolet-visible
LMCT	Ligand-to-metal charge transfer
CT	Charge transfer
LLT	Ligand–ligand transfer
MLCT	Metalto-ligand charge transfer
IVCT	Intervalence charge transfer
pK_a	Negative decadic logarithm of the acid dissociation constant (K_a)
pH	Negative decadic logarithm of the hydrogen ion concentration

Acknowledgements

I would like to thank the publisher, the Royal Society of Chemistry, for the opportunity to write this book!

Recommended Reading

Please have a look into the following references:

Transfer of polyoxometalates into the organic phase by cation modification:

1. D. Gabb, C. P. Pradeep, T. Boyd, S. G. Mitchell, H. N. Miras, D.-L. Long and L. Cronin, *Polyhedron*, 2013, **52**, 159–164.

Original publication of Bernt Krebs:

1. M. Bösing, A. Nöh, I. Loose and B. Krebs, *J. Am. Chem. Soc.*, 1998, **120**, 7252–7259.

Heteroelements in polyoxometalates:

1. J.-C. Raabe, F. Jameel, M. Stein, J. Albert and M. J. Poller, *Dalton Trans.*, 2024, **53**, 454–466.

Ligand exchange in polyoxometalates:

1. J. Breibeck, N. I. Gumerova and A. Rompel, *ACS Org. Inorg. Au*, 2022, **2**, 477–495.
2. G. Taban-Çalışkan, D. Mesquita Fernandes, J.-C. Daran, D. Agustin, F. Demirhan and R. Poli, *Inorg. Chem.*, 2012, **51**, 5931–5940.
3. E. Collange, L. Metteau, P. Richard and R. Poli, *Polyhedron*, 2004, **23**, 2605–2610.

Thermogravimetric analysis on polyoxometalates:

1. V. Sasca, M. Stefanescu and A. Popa, *J. Therm. Anal. Calorim.*, 2003, **72**, 311–322.

Acidity of polyoxometalates:

1. K. M. Reddy, N. Lingaiah, P. S. Sai Prasad and I. Suryanarayana, *J. Solution Chem.*, 2006, **35**, 407–423.
2. J. A. Dias, J. P. Osegovic and R. S. Drago, *J. Catal.*, 1999, **183**, 83–90.

References

1. J.-C. Raabe, J. Albert and M. J. Poller, *Chem. – Eur. J.*, 2022, **28**, 1–12.
2. J.-C. Raabe, L. Hombach, M. J. Poller, A. Collauto, M. M. Roessler, A. Vorholt, A. K. Beine and J. Albert, *ChemCatChem*, 2024, **16**, e202400395.
3. J.-C. Raabe, J. Aceituno Cruz, J. Albert and M. J. Poller, *Inorganics*, 2023, **11**, 138.
4. J.-C. Raabe, M. J. Poller, D. Voß and J. Albert, *ChemSusChem*, 2023, **16**, e202300072.
5. J.-C. Raabe, T. Esser, F. Jameel, M. Stein, J. Albert and M. J. Poller, *Inorg. Chem. Front.*, 2023, **10**, 4854–4868.
6. J.-C. Raabe, T. Esser, M. J. Poller and J. Albert, *Catal. Today*, 2024, 114899.
7. T. Ueda, K. Kodani, H. Ota, M. Shiro, S.-X. Guo, J. F. Boas and A. M. Bond, *Inorg. Chem.*, 2017, **56**, 3990–4001.
8. D. Gabb, C. P. Pradeep, T. Boyd, S. G. Mitchell, H. N. Miras, D.-L. Long and L. Cronin, *Polyhedron*, 2013, **52**, 159–164.
9. D. J. Sures, S. K. Sahu, P. I. Molina, A. Navrotsky and M. Nyman, *ChemistrySelect*, 2016, **1**, 1858–1862.
10. M. Nyman, T. M. Alam, F. Bonhomme, M. A. Rodriguez, C. S. Frazer and M. E. Welk, *J. Cluster Sci.*, 2006, **17**, 197–219.
11. D. J. Sures, P. I. Molina, P. Miró, L. N. Zakharov and M. Nyman, *New J. Chem.*, 2016, **40**, 928–936.
12. T. -L. Ho, *Chem. Informationsdienst*, 1975, **6**, 1–20.
13. D. Datta, *Inorg. Chem.*, 1992, **31**, 2797–2800.
14. S. Woodward, *Tetrahedron*, 2002, **58**, 1017–1050.
15. R. S. Drago and R. A. Kabler, *Inorg. Chem.*, 1972, **11**, 3144–3145.
16. S. Uchida, *Chem. Sci.*, 2019, **10**, 7670–7679.
17. M. H. Alizadeh, S. P. Harmalker, Y. Jeannin, J. Martin-Frere and M. T. Pope, *J. Am. Chem. Soc.*, 1985, **107**, 2662–2669.
18. I. Creaser, M. C. Heckel, R. J. Neitz and M. T. Pope, *Inorg. Chem.*, 1993, **32**, 1573–1578.

19. J. A. Fernández, X. López, C. Bo, C. de Graaf, E. J. Baerends and J. M. Poblet, *J. Am. Chem. Soc.*, 2007, **129**, 12244–12253.
20. P. G. Rickert, M. R. Antonio, M. A. Firestone, K.-A. Kubatko, T. Szreder, J. F. Wishart and M. L. Dietz, *J. Phys. Chem. B*, 2007, **111**, 4685–4692.
21. Y. Martinetto, B. Pégot, C. Roch-Marchal, B. Cottyn-Boitte and S. Floquet, *Eur. J. Inorg. Chem.*, 2020, **2020**, 228–247.
22. W. Zhu, W. Huang, H. Li, M. Zhang, W. Jiang, G. Chen and C. Han, *Fuel Process. Technol.*, 2011, **92**, 1842–1848.
23. H. Cruz, N. Gomes, F. Mirante, S. S. Balula, L. C. Branco and S. Gago, *ChemistrySelect*, 2020, **5**, 12266–12271.
24. S. A. Forsyth, J. M. Pringle and D. R. MacFarlane, *Aust. J. Chem.*, 2004, **57**, 113–119.
25. U. Filek, D. Mucha, M. Hunger and B. Sulikowski, *Catal. Commun.*, 2013, **30**, 19–22.
26. H. Aliyan, R. Fazaeli and N. Habibollahi, *J. Korean Chem. Soc.*, 2012, **56**, 591–596.
27. G. Li, R. Long, S. Yang and L. Zhang, *Kinet. Catal.*, 2011, **52**, 559–563.
28. P. J. Mafa, B. B. Mamba and A. T. Kuvarega, *Sep. Purif. Technol.*, 2020, **253**, 117349.
29. S. Rahut, S. S. Basu and J. K. Basu, *Chem. Commun.*, 2019, **55**, 4825–4828.
30. T. Baba and Y. Ono, *J. Phys. Chem.*, 1996, **100**, 9064–9067.
31. Y. Zhang, L. Hu, S. Zhao, N. Liu, L. Bai, J. Liu, H. Huang, Y. Liu and Z. Kang, *RSC Adv.*, 2016, **6**, 60394–60399.
32. Z.-L. Wang and H.-P. Xi, *Z. Naturforsch., B: J. Chem. Sci.*, 2012, **67**, 495–498.
33. L. Gao, Q. Sun, X. Lin, J. Qi and K. Wang, *Colloids Surf., A*, 2013, **423**, 162–169.
34. J.-C. Raabe, F. Jameel, M. Stein, J. Albert and M. J. Poller, *Dalton Trans.*, 2024, **53**, 454–466.
35. M. Witvrouw, H. Weigold, C. Pannecouque, D. Schols, E. De Clercq and G. Holan, *J. Med. Chem.*, 2000, **43**, 778–783.
36. M. Bösing, I. Loose, H. Pohlmann and B. Krebs, *Chem. – Eur. J.*, 1997, **3**, 1232–1237.
37. T. Yamase, B. Botar, E. Ishikawa and K. Fukaya, *Chem. Lett.*, 2001, **30**, 56–57.
38. C. M. Tourné, G. F. Tourné and F. Zonnevijlle, *J. Chem. Soc., Dalton Trans.*, 1991, 143–155.
39. M. Bösing, A. Nöh, I. Loose and B. Krebs, *J. Am. Chem. Soc.*, 1998, **120**, 7252–7259.
40. A. F. Holleman, E. und N. Wiberg and G. Fischer, in *Lehrbuch Der Anorganischen Chemie*, Berlin, New York, 2009.
41. L. Wang, K. Yu, B.-B. Zhou, Z.-H. Su, S. Gao, L.-L. Chu and J.-R. Liu, *Dalton Trans.*, 2014, **43**, 6070.
42. V. Kogan, Z. Aizenshtat and R. Neumann, *New J. Chem.*, 2002, **26**, 272–274.
43. I. Boldini, G. Guillemot, A. Caselli, A. Proust and E. Gallo, *Adv. Synth. Catal.*, 2010, **352**, 2365–2370.
44. C. N. Kato, I. Nakahira, R. Kasai and S. Mori, *Eur. J. Inorg. Chem.*, 2021, **2021**, 1816–1827.
45. U. Müller, in *Anorganische Strukturchemie*, Vieweg+Teubner, Wiesbaden, 2008.
46. J. Breibeck, N. I. Gumerova and A. Rompel, *ACS Org. Inorg. Au*, 2022, **2**, 477–495.
47. G. Taban-Çalışkan, D. Mesquita Fernandes, J.-C. Daran, D. Agustin, F. Demirhan and R. Poli, *Inorg. Chem.*, 2012, **51**, 5931–5940.
48. E. Collange, L. Metteau, P. Richard and R. Poli, *Polyhedron*, 2004, **23**, 2605–2610.
49. B. Weber, in *Koordinationschemie*, Springer, Berlin, Heidelberg, 2014.
50. D. Steinborn, in *Grundlagen Der Metallorganischen Komplexkatalyse*, Springer, Berlin, Heidelberg, 2019.
51. Y. Du, A. L. Rheingold and E. A. Maatta, *J. Am. Chem. Soc.*, 1992, **114**, 345–346.
52. J. B. Strong, B. S. Haggerty and A. L. Rheingold, *Chem. Commun.*, 1997, **6**, 1137–1138.
53. J. B. Strong, R. Ostrander, A. L. Rheingold and E. A. Maatta, *J. Am. Chem. Soc.*, 1994, **116**, 3601–3602.

54. K. Y. Monakhov, C. Gourlaouen, R. Pattacini and P. Braunstein, *Inorg. Chem.*, 2012, **51**, 1562–1568.
55. A. Müller, S. K. Das, V. P. Fedin, E. Krickemeyer, C. Beugholt, H. Bögge, M. Schmidtmann and B. Hauptfleisch, *Z. Anorg. Allg. Chem.*, 1999, **625**, 1187–1192.
56. V. Sasca, M. Stefanescu and A. Popa, *J. Therm. Anal. Calorim.*, 2003, **72**, 311–322.
57. P. Pyykkö and M. Atsumi, *Chem. - Eur. J.*, 2009, **15**, 186–197.
58. I. A. Weinstock, J. J. Cowan, E. M. G. Barbuzzi, H. Zeng and C. L. Hill, *J. Am. Chem. Soc.*, 1999, **121**, 4608–4617.
59. K. M. Sundaram, W. A. Neiwert, C. L. Hill and I. A. Weinstock, *Inorg. Chem.*, 2006, **45**, 958–960.
60. W. A. Neiwert, J. J. Cowan, K. I. Hardcastle, C. L. Hill and I. A. Weinstock, *Inorg. Chem.*, 2002, **41**, 6950–6952.
61. S. Himeno, M. Takamoto and T. Ueda, *Bull. Chem. Soc. Jpn.*, 2005, **78**, 1463–1468.
62. D. Nowicka, N. Vadra, E. Wieczorek-Szweda, V. Patroniak and A. Gorczyński, *Coord. Chem. Rev.*, 2024, **519**, 216091.
63. R. Contant and R. Thouvenot, *Inorg. Chim. Acta*, 1993, **212**, 41–50.
64. A. Ogawa, H. Yamato, U. Lee, H. Ichida, A. Kobayashi and Y. Sasaki, *Acta Crystallogr., Sect. C: Cryst. Struct. Commun.*, 1988, **44**, 1879–1881.
65. J. Zhang, Y. Huang, J. Hao and Y. Wei, *Inorg. Chem. Front.*, 2017, **4**, 1215–1218.
66. M. T. Pope and T. F. Scully, *Inorg. Chem.*, 1975, **14**, 953–954.
67. L. Pettersson, I. Andersson, J. H. Grate and A. Selling, *Inorg. Chem.*, 1994, **33**, 982–993.
68. C. M. Flynn and M. T. Pope, *Inorg. Chem.*, 1971, **10**, 2524–2529.
69. J. Tucher, S. Schlicht, F. Kollhoff and C. Streb, *Dalton Trans.*, 2014, **43**, 17029–17033.
70. S. Herrmann, J. T. Margraf, T. Clark and C. Streb, *Chem. Commun.*, 2015, **51**, 13702–13705.
71. D. V. Evtuguin, C. P. Neto, J. Rocha and J. D. Pedrosa de Jesus, *Appl. Catal., A*, 1998, **167**, 123–139.
72. I. Efremenko and R. Neumann, *J. Am. Chem. Soc.*, 2012, **134**, 20669–20680.
73. S. P. Khranenko, A. S. Sukhikh and S. A. Gromilov, *J. Struct. Chem.*, 2020, **61**, 293–298.
74. J. K. Lee, J. Melsheimer, S. Berndt, G. Mestl, R. Schlögl and K. Köhler, *Appl. Catal., A*, 2001, **214**, 125–148.
75. M. Nakamura, M. S. Islam, M. A. Rahman, R. N. Nahar, M. Fukuda, Y. Sekine, J. N. Beltramini, Y. Kim and S. Hayami, *RSC Adv.*, 2021, **11**, 34558–34563.
76. A. J. Bridgeman, *Chem. - Eur. J.*, 2006, **12**, 2094–2102.
77. R. Mattes, H. Bierbüsse and J. Fuchs, *Z. Anorg. Allg. Chem.*, 1971, **385**, 230–242.
78. P. Müscher-Polzin, C. Näther and W. Bensch, *Z. Naturforsch., B:J. Chem. Sci.*, 2020, **75**, 583–588.
79. NMR Lab, Available from: https://chem.ch.huji.ac.il/nmr/techniques/expts.html.
80. Y.-G. Chen, J. Gong and L.-Y. Qu, *Coord. Chem. Rev.*, 2004, **248**, 245–260.
81. M. Haouas, J. Trébosc, C. Roch-Marchal, E. Cadot, F. Taulelle and C. Martineau-Corcos, *Magn. Reson. Chem.*, 2017, **55**, 902–908.
82. N. V. Maksimchuk, G. M. Maksimov, V. Y. Evtushok, I. D. Ivanchikova, Y. A. Chesalov, R. I. Maksimovskaya, O. A. Kholdeeva, A. Solé-Daura, J. M. Poblet and J. J. Carbó, *ACS Catal.*, 2018, **8**, 9722–9737.
83. M. Pascual-Borràs, X. López, A. Rodríguez-Fortea, R. J. Errington and J. M. Poblet, *Chem. Sci.*, 2014, **5**, 2031.
84. P. A. Lorenzo-Luis, P. Martin-Zarza, A. Sánchez, C. Ruiz-Pérez, M. Hernádez-Molina, X. Solans and P. Gili, *Inorg. Chim. Acta*, 1998, **277**, 139–150.
85. P. Gili, P. A. Lorenzo-Luis, P. Martin-Zarza, S. Domínguez, A. Sánchez, J. M. Arrieta, E. Rodriguez-Castellón, J. Jiménez-Jiménez, C. Ruiz-Pérez, M. Hernádez-Molina and X. Solans, *Transition Met. Chem.*, 1999, **24**, 141–151.

86. M. Nyman, T. Rahman and I. Colliard, *Acc. Chem. Res.*, 2023, **56**, 3616–3625.
87. I. P. Gerothanassis, *Prog. Nucl. Magn. Reson. Spectrosc.*, 2010, **56**, 95–197.
88. A. Selling, I. Andersson, J. H. Grate and L. Pettersson, *Eur. J. Inorg. Chem.*, 2000, 1509–1521.
89. M. J. Poller, S. Bönisch, B. Bertleff, J.-C. Raabe, A. Görling and J. Albert, *Chem. Eng. Sci.*, 2022, **264**, 118143.
90. M. Hunger and J. Weitkamp, *Angew. Chem., Int. Ed.*, 2001, **40**, 2954–2971.
91. K. P. Barteau, J. E. Lyons, I. K. Song and M. A. Barteau, *Top. Catal.*, 2006, **41**, 55–62.
92. I. K. Song, H. S. Kim and M. S. Chun, *Korean J. Chem. Eng.*, 2003, **20**, 844–849.
93. T. Yamase, *Chem. Rev.*, 1998, **98**, 307–325.
94. H. Salavati and N. Rasouli, *Mater. Res. Bull.*, 2011, **46**, 1853–1859.
95. H. Li, L. Swenson, R. J. Doedens and M. I. Khan, *Dalton Trans.*, 2016, **45**, 16511–16518.
96. K. M. Reddy, N. Lingaiah, P. S. Sai Prasad and I. Suryanarayana, *J. Solution Chem.*, 2006, **35**, 407–423.
97. J. A. Dias, J. P. Osegovic and R. S. Drago, *J. Catal.*, 1999, **183**, 83–90.
98. I. Efremenko and R. Neumann, *J. Phys. Chem. A*, 2011, **115**, 4811–4826.

5 Catalytic Applications of Polyoxometalates

5.1 Catalysis

5.1.1 Principle

Catalysis is a key discipline in chemistry. A catalyst is a chemical substance that speed up the rate of a chemical reaction by providing an alternative reaction path with reduced activation energy. However, the Gibbs free energy of the chemical reaction ΔG is unchanged. The catalyst is therefore both the reactant (starting material) and the product of the reaction. This means that the catalyst is not consumed during the reaction.[1]

A distinction must be made between homogeneous and heterogeneous catalysis. In homogeneous catalysis, the catalyzed reaction takes place in a single phase. The catalyst is soluble in the solvent of the reaction. In heterogeneous catalysis, the catalyst forms its own phase. This includes reactions of gaseous reactants that are converted to liquid or gaseous reaction products on a solid-state catalyst material. However, it must be noted that it is more difficult to recover the homogeneous catalyst, as it must first be separated from the reactants and products after the reaction. The recovery of a heterogeneous catalyst is easier as it forms its own phase. As homogeneous catalysts can be specifically tailored for a catalytic reaction and catalytic mechanisms can be elucidated more easily, the advantages of homogeneous catalysis often outweigh the disadvantages for the user.

RSC Foundations No. 3
Polyoxometalate Chemistry
By Jan-Christian Raabe

Published by the Royal Society of Chemistry, www.rsc.org

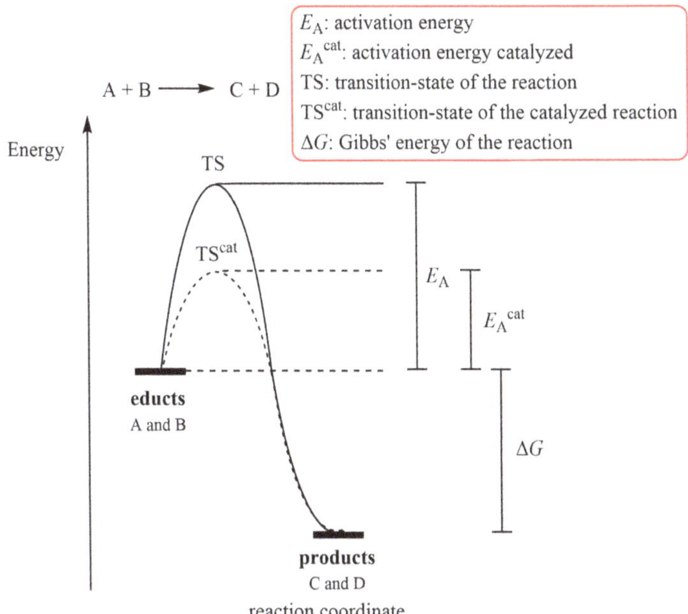

Figure 5.1 The principle of catalysis on an energy level diagram. Conversion of educts A and B to the reaction products C and D. An alternative reaction pathway is opened with reduced activation energy E_A^{cat} and an alternative transition state TS^{cat}. The Gibbs' free energy ΔG is unchanged.[1]

The principle of catalysis is shown in Figure 5.1 on a thermodynamic energy level diagram.[1]

Today, 80% of all products from the chemical industry are produced catalytically, meaning that catalysis has a fundamental economic role.[2]

The term bifunctional catalysis is used if two properties of a catalyst material are used for the catalytic mechanism. For example, a POM catalyst can be an acid and also contain active metals that can change their oxidation states. The catalyst can therefore be an acid/base and a RedOx catalyst at the same time.

5.1.2 Biomass as Substrate

Biomass is an interesting alternative raw material because it is the only available regenerative carbon source on planet earth. It includes carbohydrates and lignocellulose. Lignocellulose consists of fibrous cellulose and hemicellulose structures surrounded by lignin. For these reasons, the biomass substrate is an interesting raw material whose catalytic conversion is attracting academic and industrial interest.

The conversion of biomass includes the following steps:

1. Fractionation. Separation of the individual components (cellulose, hemicellulose and lignin).
2. Oxidative cleavage of the C–C bonds into the elementary components (monosaccharides such as glucose).
3. Further oxidation of monosaccharides and C–C splitting to small, organic acids like formic acid (FA).

FA is a very valuable platform chemical with a wide range of applications in the chemical industry. It has the following uses:[3]

1. As a C_1 building block in organic chemistry.
2. As a source of carbon monoxide (CO), which can be generated upon heating, according to $HCOOH \rightarrow CO + H_2O$.
3. As a liquid storage molecule for hydrogen (H_2) and carbon dioxide (CO_2) according to $H_2 + CO_2 \rightarrow CHOOH$ or $H_2O + CO \rightarrow CHOOH$.[4]

H_2 storage is an important technology. H_2 can be stored long term and released again if needed. For both CO and H_2, the following applications are used in industry:

- CO and H_2 can be directly used in the Fischer–Tropsch synthesis for producing longer alkane chains, which can be used as fuels.
- CO and H_2 can be used for the methanol production.
- N_2 and H_2 are used in the Haber–Bosch process for NH_3 production (5% of global natural gas consumption and 2% of global energy).

A result of the over-oxidation of biomass-based substrates is the unwanted greenhouse gas carbon dioxide CO_2. CO_2 is the thermodynamically preferred oxidation end-product as shown in Figure 5.2.

The development of a catalyst that promotes the oxidation of biomass to the desired short-chain organic acids and prevents over-oxidation to CO_2 is required. However, in the field of green and sustainable chemistry, it is desirable to use water as a solvent and molecular O_2 as an oxidizing agent. Therefore, the selected catalyst system must also be soluble and stable in water. POMs are suitable catalysts for this purpose, because they are prepared and can be stored in aqueous media. It is also possible to prepare POMs on a large scale, meaning that a promising catalytic approach can be implemented into industrial use. Moreover, POMs can easily be tuned for a desired catalytic application, due to the choice of the framework and foreign element.

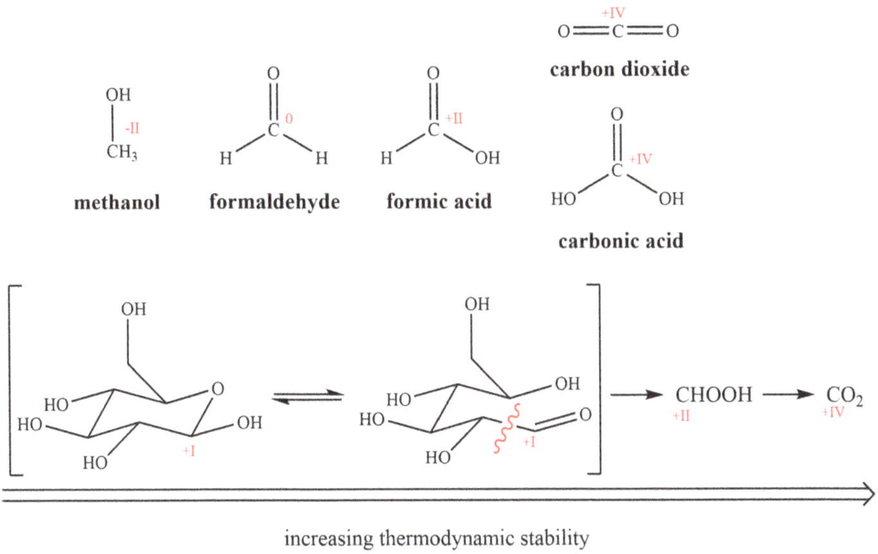
+IV
O=C=O
carbon dioxide
OH
-II
CH3
methanol
O
0
C
H H
formaldehyde
O
+II
C
H OH
formic acid
O
+IV
C
HO OH
carbonic acid
OH
HO
HO
O
OH
HO
+I
OH
HO
HO
OH
O
HO
+I
CHOOH
+II
CO2
+IV
increasing thermodynamic stability

Figure 5.2 Oxidation states of carbon in different organic compounds. In CO_2, carbon has its highest oxidation state of +4 and is therefore the thermodynamically preferred oxidation product.

5.1.3 Photocatalysts and Enzymes

A photocatalyst is a chemical substance that changes into an excited state if it is exposed to electromagnetic radiation (IR, UV or visible range) and thus catalyzes a chemical reaction. The phenomenon of photocatalysis means that a chemical reaction is accelerated or initiated in the presence of a photocatalyst under the influence of electromagnetic radiation. Photons ($h\nu$) do *not* act as catalysts, as photons are absorbed and consequently consumed by the photocatalyst during photocatalysis.[1]

Enzymes are proteins that can catalyze chemical reactions. In biochemistry, enzymes play a significant role in vital processes. Like all proteins, enzymes consist of a chain of amino acids that are linked together by a peptide bond. All proteins differ in length and in the sequence of amino acids. There are 20 amino acids that make up biological proteins, the so-called proteinogenic amino acids. Each protein forms a defined three-dimensional structure that is relevant for a respective biological function. Each catalytically active protein (enzyme) has a so-called active center into which a defined substrate fits, which is coordinated by hydrogen bonds or by coordinative interactions to a metal, which has been complexed by the amino acid chain in a separate step. After the catalytic reaction in the active center is done, the reaction products are released. In addition to the specific substrate, other molecules can also fit into the active center but cannot be converted by the enzyme. This is called inhibition (poisoning) of the enzyme, meaning that the enzyme can no longer fulfill its biological (catalytic) function. If an enzyme is exposed to the substrate and the inhibitor, the substrate and inhibitor will compete for the active center.[5]

5.1.4 Catalytic Cycle

A catalytic cycle is a multi-step reaction mechanism involving a catalyst. Due to the regeneration of the catalyst, the steps are visualized as a cycle. In general, there are at least two steps involved: (1) the coordination of the substrates onto the catalytic species, and (2) release of the products under regeneration of the catalyst.[1]

5.2 Reaction Control of Homogeneous Catalytic Processes

Homogeneous catalytic processes can be done in the following ways:

- Monophasic in aqueous or organic media. This requires that the catalyst, substrate and products are soluble in the same phase.
- Biphasic with an aqueous and an organic phase (two immiscible solvents).
  - Reaction as *in situ* extraction. The catalyst is soluble in phase A (aqueous or organic) and the substrates or products are soluble in phase B (organic or aqueous). Phase A, in which the catalytic cycle takes place, is called the catalyst phase. The products are more soluble in phase B and are therefore directly extracted into phase B. Direct extraction shifts the equilibrium of the chemical reaction to the product side, as the product is continuously removed from the catalyst phase (Le Chatelier's principle). Example 1: The catalyst is water soluble (the catalytic cycle takes place in the aqueous phase). The reaction product is directly extracted into the organic phase, as it is more soluble in organic solvents. Example 2: The catalyst is soluble in water and the substrate is in the organic phase (and not soluble in aqueous media). For the catalytic conversion, the substrate is transferred into the aqueous phase, where it is converted to a water-soluble product. The reaction product remains in the aqueous catalyst phase after the reaction is done.[6]

- Reaction in a thermomorphic solvent system. The reaction takes place in two solvents, which are not miscible under ambient conditions (*e.g.* water and 1-butanol). For example, the catalyst is soluble in aqueous media and the substrates and/or products are not. This means that the aqueous phase contains the catalyst, and the organic phase contains the substrates. Under certain reaction conditions (increased temperature and/or pressure), both solvents are miscible and form a single phase in which the catalytic reaction takes place. After the reaction has ended and the mixture is brought under ambient conditions, both phases are separated again. The aqueous phase contains the catalyst and the organic phase now contains the reaction products.[7]

The advantages of performing catalytic reactions in biphasic media are:

- easy separation of products and catalyst (by phase separating);
- recycling and isolating of the fully intact catalyst (by phase separating);
- reuse of the catalyst in a new cycle or application (by phase separating); and
- shift of the chemical equilibrium.

The following problems can occur:

- Volatile organic solvents can pose a health hazard.
- Explosive mixtures of organic solvents and molecular O_2 can form (O_2 is always used as oxidant for oxidative conversions). This problem can be reduced by:
  - diluting the gas phase with an inert gas (*e.g.* N_2 or Ar) to stay under the lower explosive limit;
  - increasing the O_2 pressure to stay over the upper explosive limit.

The following adjustments can be made to improve the mixing of the two phases:

- The solubility of the catalyst can be adjusted by the choice of cation (for POMs, inorganic cations are chosen for solubility in aqueous media and organic cations for organic media).
- The solubility of the substrates can be adjusted by the choice of suitable groups (pure organic substrates are only soluble in

organic media, hydrophilic groups like SO_3^- mediate substrate solubility in aqueous media). Introduction of hydrophilic groups (such as SO_3^-) into organic molecules ensures that they become water-soluble. In this way, substrate solubility can be specifically influenced.

- The mixing of the two phases can be increased by adding so-called dispersing agents (*e.g.* by adding surfactants).

Biphasic reactions are often preferred to monophasic reactions in order to utilize the advantages of heterogeneous catalysis with the aim of recovering the fully intact catalyst. A summary of different approaches can be found in Figure 5.3.

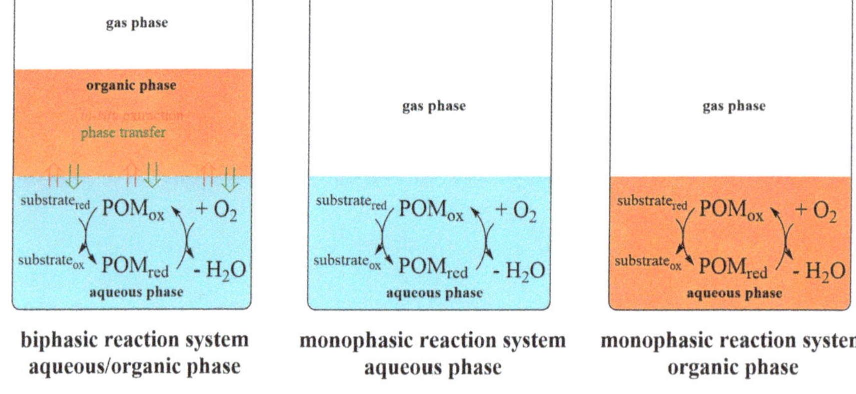

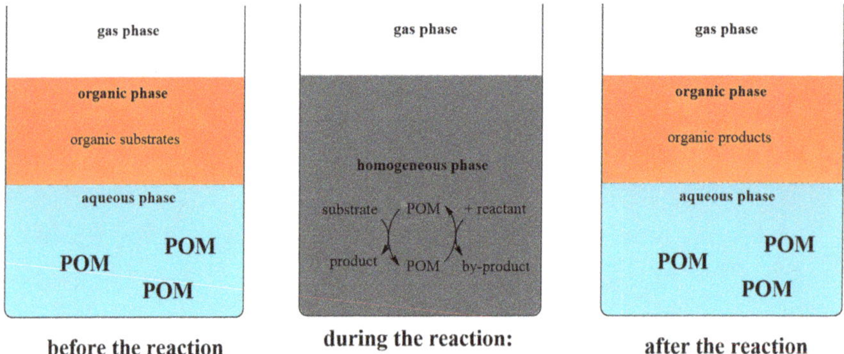

Figure 5.3 Homogeneous catalyzed reactions can be performed monophasic (in aqueous or organic media) or biphasic (as *in situ* extraction or with a thermomorphic solvent system). The different options have different advantages and disadvantages.

5.3 Acid/Base Catalysis

POM acids can act as acid or base catalysts and are stronger acids than inorganic mineral acids. Using POMs, unwanted side reactions can be avoided such as sulfonation, chlorination or nitration.[6,8,9]

POM acids can catalyze esterification reactions as shown in Scheme 5.1. Those esterification reactions can be done in a biphasic medium (*e.g.* FA with 1-hexanol/1-heptanol) or in a single phase (*e.g.* FA in methanol to form methyl formate).[6,10–13]

Another type of acid catalyzed reaction is a dehydration. Cellulose can be depolymerized to its monosaccharides (glucose). The glucose can be dehydrated to small organic acids (*e.g.* levulinic acid, LA, and FA). However, there is a side reaction, in which insoluble, dark-colored humins are formed. Humins are completely insoluble organic waste polymers. Dissolution is only possible in concentrated NaOH under destruction of the polymer chains. Normally, humins are disposed of, but valuable carbon that could have been converted into FA or other high-value products is also lost. Those humins are formed from C6 sugars, hydroxymethylfurfural (HMF) or from similar intermediates. This has motivated recent research to search for catalytic conversions of humins into valuable products instead of disposing of them.[14–19] A possible reaction pathway and a possible fragment of a humin structure are shown in Figure 5.4. Note that the real humin structure is much more complex.[6]

POM acid catalyzed dehydrations can also be done in deep eutectic solvents (DES). A DES is a mixture of compound A and B, whose melting point at the eutectic point is lower than that of the pure substances. The eutectic point refers to a special composition of two components A and B. Here, the entire mixture melts without forming a melt consisting of different phases. An example of an application is the dehydration of fructose to HMF, using tetraethyl ammonium chloride (hydrogen bond acceptor, HBA, compound A) and levulinic

$$R^1COOH + HOCH_2R^2 \xrightleftharpoons[+/-\ H_2O]{H_{3+x}[PV_xMo_{12-x}O_{40}]} R^1COOCH_2R^2$$

Scheme 5.1 POM acid catalyzed esterification. An organic acid (R^1COOH) is converted under POM acid catalysis $H_{3+x}[PV_xMo_{12-x}O_{40}]$ with an organic alcohol (R^2CH_2OH) to an organic ester ($R^1COOCH_2R^2$).

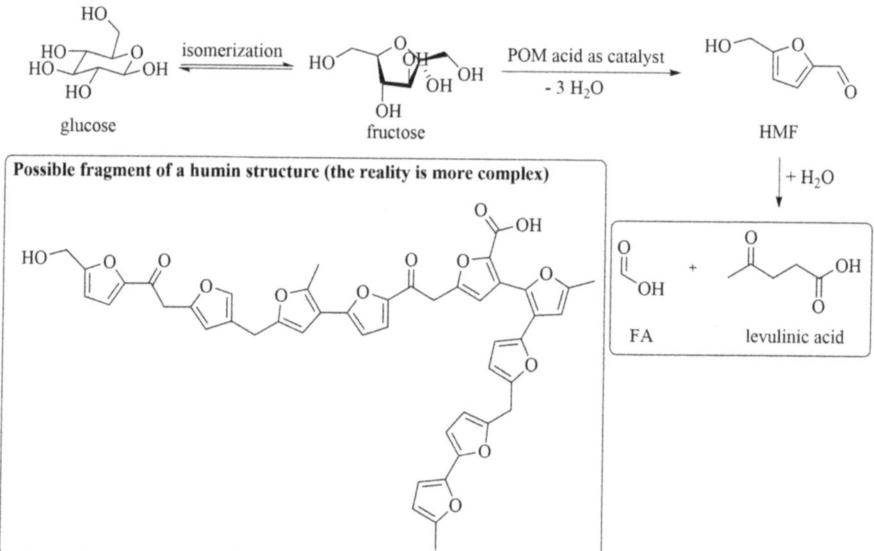

Figure 5.4 Building blocks for humin formation and conversion of glucose to levulinic acid with HMF as an intermediate. A possible fragment of a humin structure is shown on the bottom.[12]

acid (hydrogen bond donor, HBD, compound B) in a molar ratio of 1 : 2 as a deep eutectic solvent system (see Figure 5.5).[6,20]

Another acid catalyzed dehydration is known for Keggin-type W based clusters with a molecular stoichiometry of $H_{8-n}[X^{n+}W_{12}O_{40}]$. X can be P, Si, Al or Co. Here, 2-butanol is dehydrated to *cis*/*trans*-2-butene. The catalytic cycle is shown in Figure 5.6. The reaction rates depend on the rate constant k_2 for the C–O cleavage and K_4 the reaction constant of the equilibrium for dimer formation. At low 2-butanol pressures (<0.1 kPa), the dehydration rate decreases with $H_3[PW_{12}O_{40}] > H_4[SiW_{12}O_{40}] > H_5[AlW_{12}O_{40}] > H_6[CoW_{12}O_{40}]$. A reversed

HO, HO, OH, OH, O, OH

DES*
POM acid
- 3 H_2O

HO, O, O

froctose

HMF

*DES mixture:

N, Cl, HO, O, O

tetraethyl ammonium chloride

levulinic acid

Figure 5.5 Dehydration of fructose to HMF with a POM acid catalyst in a DES mixture, consisting of tetraethyl ammonium chloride and levulinic acid.[20]

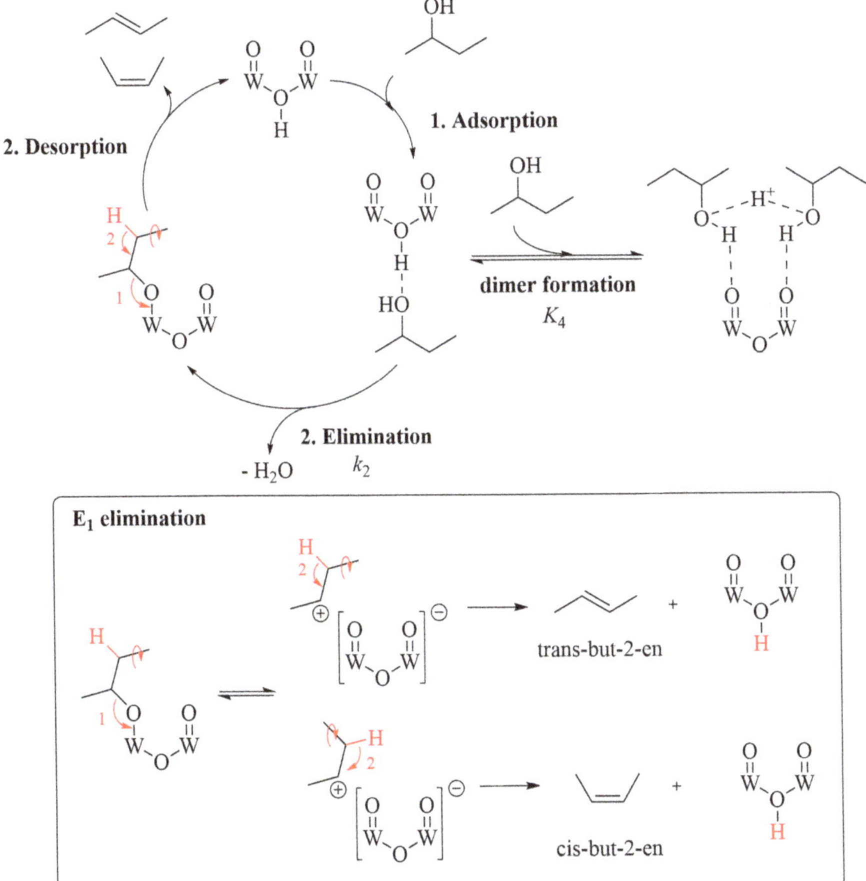

Figure 5.6 Catalytic cycle of 2-butanol dehydration, catalyzed by W based clusters.[12,21]

trend was observed at higher pressures. $H_3[PW_{12}O_{40}]$ is the strongest acid in this sequence. For $H_6[CoW_{12}O_{40}]$ the anionic charge is six, meaning that the basicity of the anion is the highest. It is possible that the mechanism benefits from stronger acids at lower pressures and from weaker acids at higher pressures. The elimination takes place according to the E_1 mechanism, forming a carbenium cation, which is associated with the POM anion. In the last step of the E_1 sequence, the kinetic *cis* or the thermodynamically preferred *trans* product can be formed, due to the rotatability of the C–C single bonds.[12,21]

Base catalysis with POMs is less explored compared to acid catalysis. Highly negatively charged POM anions are suitable candidates for base catalysis, because they have electron-rich oxygen atoms on their surfaces, which can act as basic active sites. An attack on electrophiles is possible. Lacunary-type anions are known to act as base catalysts, *e.g.* the anion $[SiW_{10}O_{36}]^{8-}$. The negative charge can be increased by generating more vacancies in the lacunary-type structure, making lacunary-type anions suitable candidates for controlling the base activity. Examples for base catalyzed reactions by POMs are the Knoevenagel condensation and the cyanosilylation with trimethylsilyl cyanide (TMSCN), as shown in Figure 5.7.[12,22]

In the Knoevenagel reaction, the C–H acidity of malononitrile is used, which attacks the aldehyde. Water is eliminated in order to form

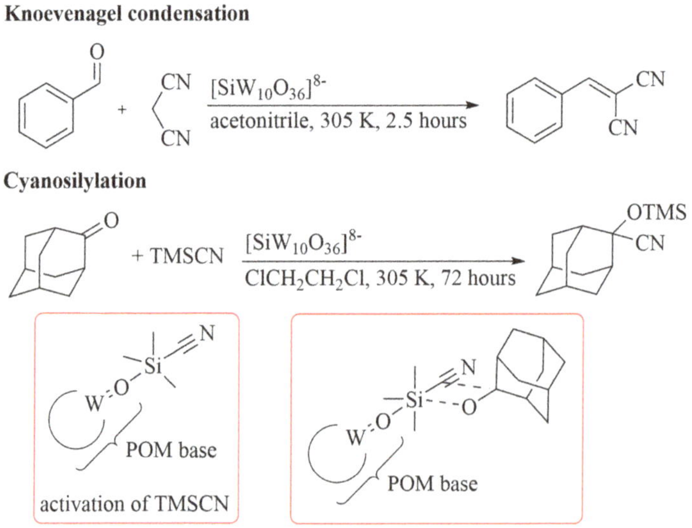

Figure 5.7 Base catalysis with the lacunary-type structure $[SiW_{10}O_{36}]^{8-}$: Knoevenagel condensation and Cyanosilylation. For the Cyanosilylation, the terminal oxo ligand coordinates to the Si atom of TMSCN, increasing the nucleophilicity of the CN group.

a C=C double bond. Cyanosilylation takes place by coordination of a terminal oxo ligand of the POM base to the Si atom of TMSCN, which acts as a Lewis acid. This weakens the Si–CN bond. The oxygen atom of the ketone can also coordinate to Si, whereby the carbonyl carbon atom is positively polarized. Now, the CN group is transferred to the carbonyl carbon atom, which strengthens the bond between the carbonyl oxygen atom and the Si atom. In the final step, the POM dissociates, regenerating the active catalyst.[12,22]

5.4 Reduction and Oxidation Catalysis

The general principle of RedOx catalysis is based on the following steps:

1. $POM_{ox} + substrates_{red} \rightarrow POM_{red} + substrates_{ox}$ (substrate oxidation step)
2. $POM_{red} + O_2 \rightarrow POMox + H_2O$ (catalyst reoxidation step)

In the first step, the POM catalyst is in its fully oxidized state (all metals are present in their highest oxidation states). The substrates are present in their reduced form (*e.g.* in glucose the carbon atoms are in the oxidation state of +1). An electron transfer takes place from the substrates to the POM, reducing the POM and forming the reduced POM_{red} species. In parallel, the substrates are oxidized, forming the oxidation products, $substrates_{ox}$. This step is called the substrate oxidation step. In the second step, the reduced POM_{red} species is oxidized again by molecular O_2, forming the species POM_{ox} and one molecule of water. This step is called the catalyst reoxidation step. For example, reducing the anion $\left[PMo_{12}^{+VI}O_{40}\right]^{3-}$ with two electrons, results in the species $\left[PMo_2^{+V}Mo_{10}^{+VI}O_{40}\right]^{5-}$ with an anionic charge increased by two. The increased anionic charge can be compensated by two protons and, after reoxidation with O_2, one molecule of water is formed. Different examples are known for RedOx catalysis with POMs, *e.g.* the delignification or the oxidation of biomass-based substrates.[6]

5.4.1 Oxidation Catalysis Using Biomass-based Substrates

5.4.1.1 Delignification

The pulp-and-paper industry is a very big global industry. It transforms wood into chemical pulp, a fiber-based material, which can be used to produce paper. For this aim, two processes are needed: pulping and bleaching. Pulping is required to destroy and eliminate the lignin from the wood, a step that is called delignification.[6,23]

1. Pulping process: Wood is pulped and delignified.[6,23]
2. Bleaching process: Oxidation step to acquire the needed pulp brightness.[6,23]

In order to avoid toxic sulfur and chlorine-based chemicals, POMs are ideal and regenerable oxidizing catalysts for the delignification of lignocellulose.[6,23]

It has been shown that selective delignification is possible due to the lower lignin oxidation potential in comparison to polysaccharides.[6,24,25] The POM acid $H_8[PV_5Mo_7O_{40}]$ is very promising, because this compound can oxidize phenolic and non-phenolic lignin structures, as shown in Figure 5.8.[26–28] In general, a model substance is defined if a similar and structurally simplified motif is catalytically converted in order to draw conclusions about the catalytic conversion of real and significantly more complex substrates. In this example, the phenol ether-based lignin structure is extremely complex, so the catalytic conversion is first tested on simplified molecular phenol ethers in order to compare the results of the catalytic conversion on real lignins later. Under aerobic conditions, the thermodynamic condition can be written as follows: the oxidation potential of the substrate $E^0_{\text{substrate}}$ needs to be lower than the potential of the catalyst E^0_{catalyst}. E^0_{catalyst} must be lower than the potential of the reduction of molecular O_2 $E^0_{\text{oxygen}} = +1.23$ V (at pH 1).[28] This condition can be written as: $E^0_{\text{substrate}} < E^0_{\text{catalyst}} < E^0_{\text{oxygen}}$.[6,28]

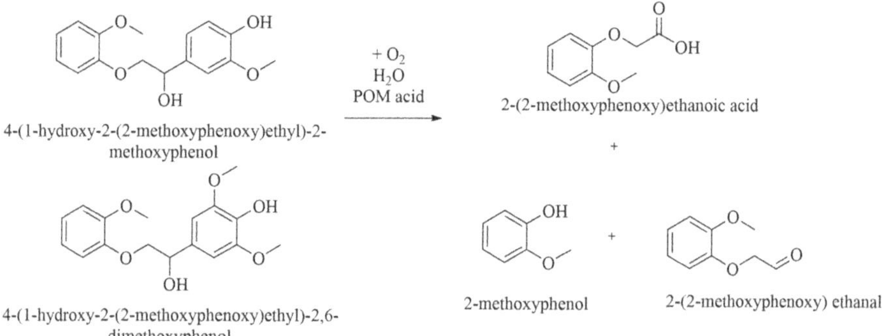

Figure 5.8 Catalytic oxidative delignification of two phenolic ether-based model substances to different products.

Delignification requires very low pH values.[29] In addition to water as solvent, the use of molecular O_2 makes this an example of a green chemistry process.[6]

The potential of oxidative conversion of phenolic ethers is much higher than the destruction of nonphenolic ethers.[6,30,31] Different mechanisms for the catalytic destruction were discussed, but for the conversion of vanillyl and veratryl alcohol, the results indicated a one-electron oxidation step of the substrate by the POM catalyst.[6,32,33]

Due to the diversity of this research field, delignification remains an open research project with many other unanswered questions.[34–41]

5.4.1.2 Generating Industrial Value Products

For the catalytic conversion of biomass, a bifunctional catalysis is required. In the first step, the biomacromolecules are split into their monomers by acid/base catalysis. In the second step, the monomers are oxidized to short-chain carboxylic acids (like FA). This principle is of fundamental significance for the following chapter.

Biomass oxidation is implemented in the so-called OxFA process (oxidative formation of FA from biomass), which is commercially employed by OxFA GmbH and produces FA from fossil raw materials.[42,43] The OxFA process was investigated for the V(v) substituted Keggin-type POMs $H_{3+x}[PV_xMo_{12-x}O_{40}]$. All POMs with $x = 0$ to 6 are catalytically active and can convert glucose to FA with CO_2 as a by-product (Figure 5.9). POMs $H_3[PMo_{12}O_{40}]$ and $H_4[PVMo_{11}O_{40}]$ show a

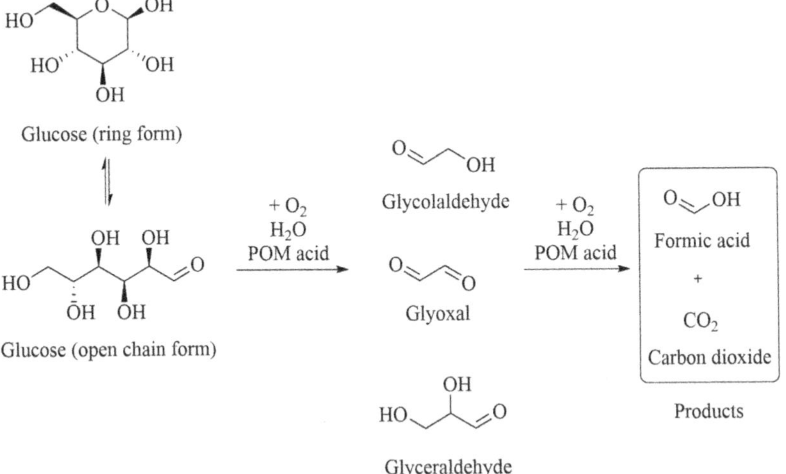

Figure 5.9 POM catalyzed conversion of glucose to FA and/or CO_2 as by-product.

very low activity and the reoxidation step $HPA_{red} \rightarrow HPA_{ox}$ is very slow. With higher degrees of V(v) substitution, the catalytic activity increases. However, the POMs also lose stability as the degree of substitution increases, resulting in a sharp drop in catalytic activity for $x = 6$. This is explained due to the increasing negative charge with increasing substitution degree.[44]

A possible postulated mechanism for the destruction of glucose is shown in Figure 5.10, using a hypothetical glucose dimer as an example, in order to present the concept of bifunctional catalysis. In the first step (hydrolysis), only acid/base catalytic properties are required. The second step (oxidation) requires only the RedOx activity of the metals, meaning that the metals are reduced first and oxidized in the last step by molecular O_2. The fully oxidized POM anion is regenerated.

- In the first step (the so-called hydrolysis), the POM acid (with all metals in their oxidized form) protonates the glycosidic oxygen atom (on C1), called O_g. This oxygen atom connects both glucose rings A and B.
- One molecule of water splits the dimer (hydrolysis), resulting in one glucose and another protonated glucose molecule. The proton is split off, forming two single glucose molecules and the regenerated POM acid.

The oxidation states of the metals in the POM anion do not change in this step – all metals remain in their highest oxidation states.

- In the second step, the metals of the POM anion are reduced by removing electrons from the glucose molecules in the open chain form. There are different positions where the molecule can be split. One possible position is between the second and third carbon atoms of glucose (Figure 5.10). The two binding electrons can be removed by two POM anions, resulting in two reduced POM anions (with an anionic charge increased by one), a glyoxal and a (2*R*,3*R*)-2,3,4-trihydroxybutanal molecule.
- Two water molecules can be added to both aldehyde functionalities of glyoxal. The two binding electrons between both carbon atoms can be removed by two POM anions, resulting in two reduced POM anions and two molecules of FA.
- The (2*R*,3*R*)-2,3,4-trihydroxybutanal molecule can now react according to similar pathways. Under certain circumstances, a

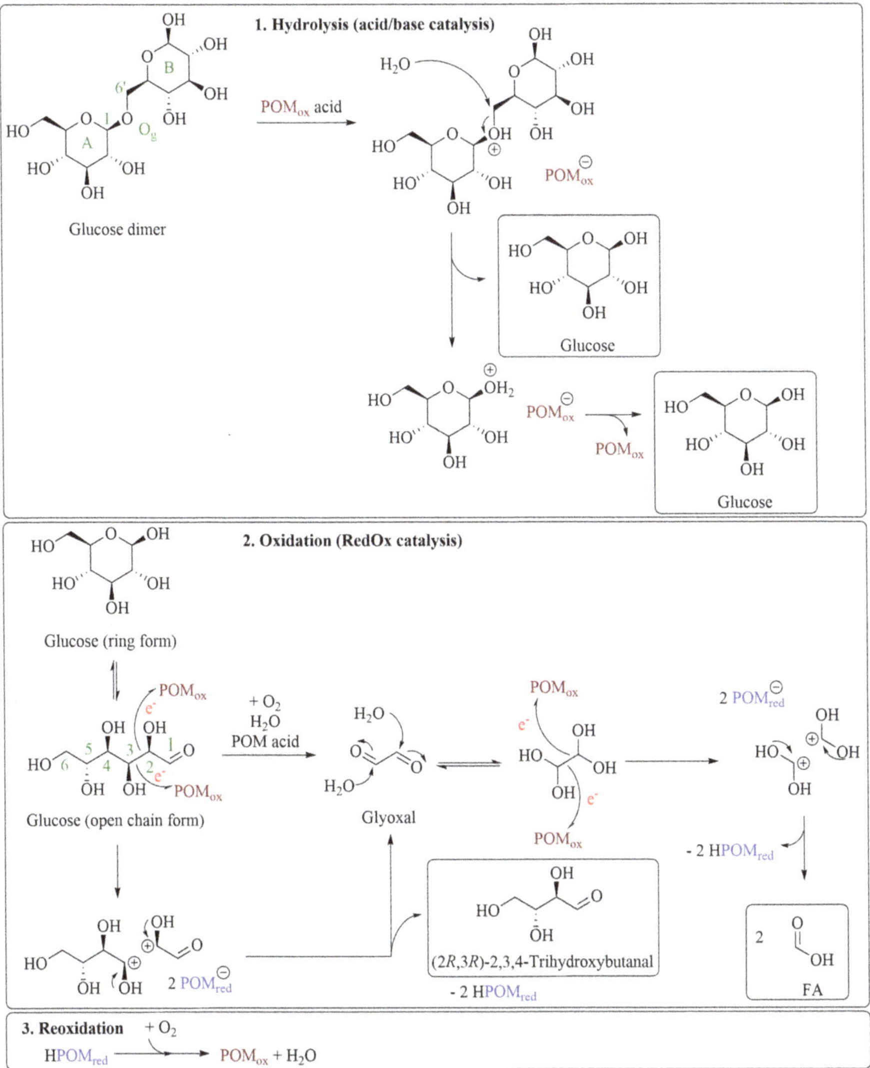

Figure 5.10 Possible reaction pathways for the catalytic (POM acid catalyzed) oxidative conversion of a hypothetical glucose dimer, explaining the formation of FA and glyoxal. There are three steps: (1) the hydrolysis of the dimer, (2) the oxidation of glucose in order to form glyoxal and FA, and (3) the reoxidation of the reduced POM species in order to regenerate the POM catalyst. Note: POM_{ox} stands for a protonated POM acid. POM_{ox}^{-} means a POM anion deprotonated by one proton.

direct oxidation of the glyoxal to oxalic acid (OA) would also be conceivable. These options underline how complex the true nature of the catalytic conversion of biomass-based substrates can be.

In this step, only RedOx catalysis takes place. The acid/base properties of the POM are not required. Note: the reaction sequence shown is only an excerpt from a possible reaction sequence. Many other pathways that can lead to different cleavage products are also possible, which makes postulating an exact reaction mechanism extremely complicated. However, it must be noted that many reaction mechanisms can be proposed on paper, potentially leading to different reaction products. Therefore, experimental analyses are necessary to verify which products are actually formed in order to assess the credibility of the proposed reaction pathways.

In the third and final step, the reduced POM anions are reoxidized by molecular O_2, forming water as a by-product and the regenerated and fully intact POM acid catalyst.

The mechanism of the catalytic conversion of a glucose dimer shown here demonstrates how complicated it is to formulate a reaction mechanism for such conversions. It is therefore clear that it seems almost impossible to formulate reaction pathways for even more complex biomass-based substrates. For a very simple substrate such as glucose, there are already different reaction pathways that can explain the formation of the found reaction products. It is likely that the individual steps of acid/base and RedOx catalysis take place in parallel. Glucose can also be split between the third and fourth carbon atom, which could explain the formation of longer-chain reaction products such as glyceraldehyde. The discussed example highlights the principle of bifunctional catalysis, which is found in the catalytic conversion of biomass-based substrates.

Furthermore, a POM anion can also accept two electrons and can be reduced by two units. With each reduction of a POM anion, the anionic charge increases by one, creating space in the cation sphere for another proton, split off from a water or a substrate molecule. After reoxidation, the anion returns to its original charge, using the excess protons to form water molecules.

Glucose is a very simple substrate to study, because the solubility in water is very high.[45] Using more complex substrates like lignocellulosic or algae-based substrates, solubility problems of those substrates in aqueous media can occur. For this purpose, *para*-toluene sulfonic acid (pTSA) can be used, acting as a solubility promoter in order to help dissolve those substrates (Scheme 5.2).[6,15,46–56]

For the OxFA process, a biphasic reaction system was developed in order to make sure that the FA is extracted during the reaction, see Figure 5.11. This reaction step is called *in situ* extraction. Water and 1-hexanol were combined, two solvents which are not

para-toluene sulfonic acid

Scheme 5.2 Molecular structure of *para*-toluene sulfonic acid.

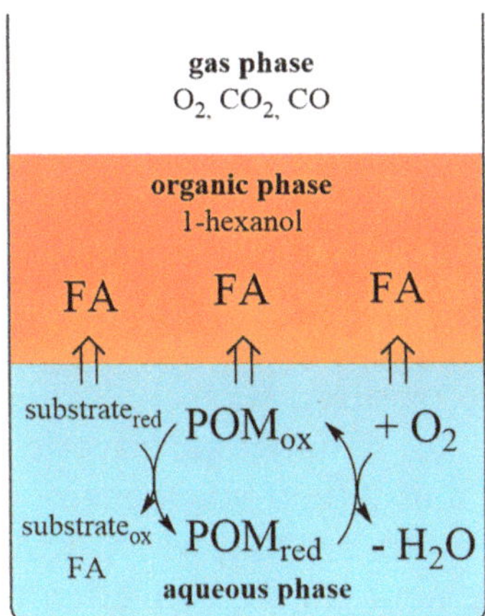

Figure 5.11 Biphasic reaction system for the catalytic conversion of biomass-based substrates to FA. FA is directly extracted into the organic 1-hexanol phase (*in situ* extraction).

miscible under ambient conditions. The POM catalyst catalyzes the glucose conversion in the aqueous phase, while the FA gets extracted into the organic 1-hexanol phase. A POM acid also catalyzes the esterification of FA with the alcohol. The hexyl ester is directly extracted in the organic phase. The hydrolysis takes place in another reactive distillation step, ensuring that FA can be collected in the distillate. Using the biphasic reaction system, up to 85% of FA can be isolated, while a maximum of 53% FA was obtained for the monophasic aqueous system under the same reaction conditions.[10]

Lignocellulosic biomass is a renewable feedstock that contains many organic functional groups, giving it the potential to replace petroleum as a feedstock for the production of carbon-based fuels. Using ionic liquids (ILs), lignocellulosic biomass can be fractionated and the biomass fragments can be oxidized with a POM catalyst. This concept is called the "POM-ionosolv concept". Using triethylammonium hydrogen sulfate (TEA)(HSO4), hemicellulose and lignin components are dissolved in the IL and the POM oxidizes the dissolved components to FA or AA.[57,58]

5.4.2 Oxidation Catalysis Using Other Substrates

5.4.2.1 Desulfurization of Fuels

Crude oil contains a mixture of different sulfur- and nitrogen-containing organic compounds. A complete removal of those organic compounds is very challenging but necessary, because after combustion they form toxic sulfur or nitrogen oxides.[59–66]

POM catalysts can catalyze the oxidation of those organic compounds and can remove them from fuels using a biphasic approach, which is known as the "extraction-coupled oxidative desulfurization" (ECODS) process. Comparable to the OxFA process, $H_8[PV_5Mo_7O_{40}]$ is used as a catalyst. This technology belongs to the class of oxidative desulfurization (ODS). With POM catalysts, the sulfur- and nitrogen-containing organic compounds like benzothiophene or dibenzothiophene are oxidized with molecular O_2 to small, water-soluble molecules, like sulfate, sulfoacetic acid (SAA), 2-sulfobenzoic acid (2-SBA) and 2-(sulfooxy)benzoic acid (2-SOBA). 2-SBA can be directly oxidized to 2-SOBA, which hydrolyzes to H_2SO_4 and salicylic acid. Salicylic acid can be converted to CO_2, FA and AA, as shown in Figure 5.12.[59]

The ECODS process is performed in a biphasic form. Here, the POM catalyst is in the aqueous phase while the fuel, which contains the sulfur and nitrogen impurities, forms a separate phase. The heterocycles are transferred to the aqueous phase for catalytic conversion, where they are converted to water-soluble products that remain in the aqueous phase with the catalyst (see Figure 5.12). A splitting of the sulfur–carbon bond takes place between S1 and C2 and/or C7a for benzothiophene and between S1 and C1a and/or C9a for dibenzothiophene.[59]

Using $H_8[PV_5Mo_7O_{40}]$ as catalyst material, a catalyst deactivation can be observed, due to the accumulation of sulfur compounds in the aqueous phase. Those polar products lead to the deactivation of the catalyst. With increasing catalytic cycles, more sulfur-containing acids accumulate in the aqueous phase, resulting in a decrease in the pH value. AA and FA show a strong deactivating effect, which can be counteracted by the addition of OA. OA can reduce V(V) to V(IV), while it is oxidized to CO_2.[67] To overcome this problem, a nanofiltration approach can be used to separate the water-soluble organic sulfur compounds from the catalyst.[6,62,63] The rejection of the POM catalyst is 99.9%. In combination with nanofiltration, iso-octane solutions (containing benzothiophene impurities) can be desulfurized after six cycles.[63]

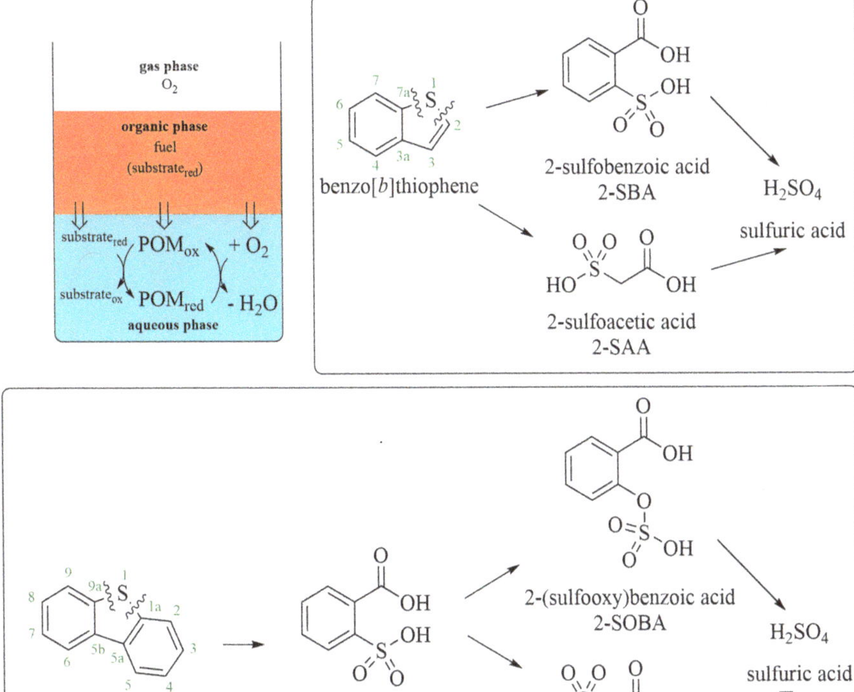

Figure 5.12 Biphasic reaction system of the ECODS process (*in situ* extraction) and the different water-soluble products formed after successful desulfurization.

AA, FA and OA can bind to the terminal oxo ligands of the POM by hydrogen bridges. The complexes that are formed after binding AA or FA to the POM inhibit and deactivate the POM catalyst, as shown in Figure 5.13. Here, AA and FA bind to two directly adjacent V(v) centers, which are the catalytic active species in the catalytic cycle. This means that those complexes cannot take part in the catalytic cycle, because they are not able to split their terminal oxo ligand. Note: There are two directly adjacent positions in a Keggin-type structure: those where the VO_6 octahedra are connected *via* shared corners or edges. Depending on the type of connection, the distance between the two catalytically active centers is either 3.35 or 3.76 Å. OA also binds *via* hydrogen bridges to the terminal oxo ligands and is able to donate two electrons to the POM, reducing two neighboring V(v) centers – each with one electron – while being oxidized to two molecules of CO_2.

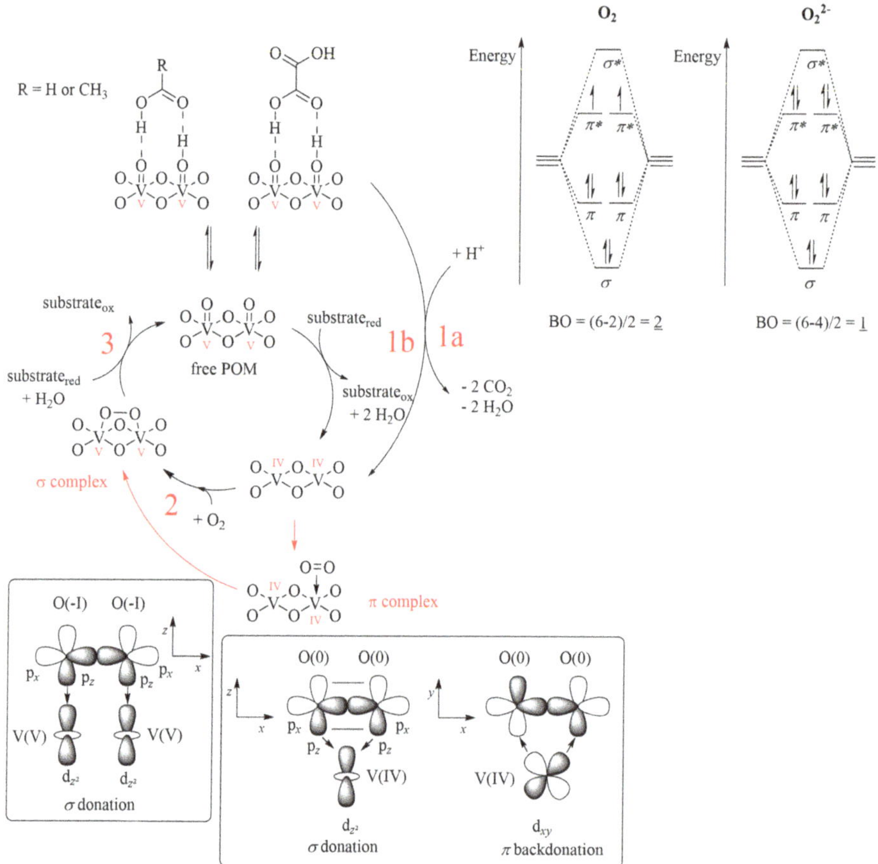

Figure 5.13 The catalytic cycle is done with Keggin-type POM acids like H_{3+x} $[PV_xMo_{12-x}O_{40}]$. Deactivation of the catalytic cycle by inhibition with AA and/or FA. AA and FA bind *via* hydrogen bonds to the terminal oxo ligands of the POM (here to two directly adjacent V(V) centers, which are the catalytic active species in the catalytic cycle). Those complexes are unable to dissociate the terminal oxo ligands, take up oxygen, and oxidize the substrate. OA is also able to bind *via* hydrogen bonds to the terminal oxo ligands. However, one electron is transfered to each V(V) center. V(V) is reduced to V(IV). Two molecules of CO_2 are formed. Four protons protonate the two terminal oxo ligands and form two molecules of water. Two free coordination sites have been formed on each V(IV) center. The catalytic cycle has been activated. Without OA the two V(V) centers can be reduced by the substrates, which are added in their reduced form substrate$_{red}$ (forming substrate$_{ox}$). The catalytic cycle is activated. Now, the two V(V) centers have been reduced in an one electron reduction step and the two terminal oxo ligands are dissociated. In the next step the regular catalytic cycle starts: formation of the π and peroxo complex (σ complex). During the transition from the π to the σ complex, the V(IV) centres are reoxidized to V(V). The substrate gets oxidized and each of the free V(V) centres associate a water molecule and regenerate the terminal oxo ligands. After this step, a fully re-generated POM acid is formed, which can either re-enter the catalytic cycle or be captured by AA/FA molecules.

Both terminal oxo ligands are cleaved off as water through protonation by the protons that serve to balance the charge of the POM anion (step 1a). This step activates the POM. Now, two V(IV) species are formed – each with a free coordination site for a new ligand (here for molecular oxygen). The V(IV) centers can be oxidized again by O_2, forming a peroxo species (step 2). The peroxo ligand coordinates to both reoxidized V(V) centers, each *via* a rotationally symmetrical σ bond in a bridging bond motif (σ complex). With the peroxo species, the substrate can be oxidized and two water molecules are needed to regenerate the terminal oxo ligands (step 3). Without OA, the V(V) centers can only be reduced by the substrate (step 1b). This step is slow compared to the reduction by OA. Here, the substrate in its reduced form ($substrate_{red}$) transfers two electrons – one to each V(V) center – resulting in an oxidation of the substrate ($substrate_{ox}$). Just like in activation with OA, two neighboring V(IV) centers are formed with a vacant coordination site for the peroxo ligand.[6,67]

The formation of the peroxo σ complex can be interpreted with a π complex of molecular O_2 in the first step. Here, the O_2 ligand forms a σ bond with a σ-molecular orbital (MO) of O_2 and the d_{z^2} orbital of V(IV). Electron density is transferred from O_2 to V(IV). V(IV) has one electron in the d orbital available (d^1 configuration), *i.e.* in the d_{xy} orbital. This orbital can now overlap with the antibonding π^*-MO of O_2 (π backdonation), forming a π bond. Electron density is transferred from V(IV) to O_2. Due to the transfer of electron density into the antibonding MOs of O_2, the bond between both oxygen atoms is weakened, resulting in a longer distance between both oxygen atoms. The bond order (BO) of O_2 is reduced to one, resulting in a σ bond between both oxygen atoms. A peroxo ligand is formed with only a σ bond between both oxygen atoms and the metal. For a better visualization, the MO schemata of O_2 and the peroxo ligand O_2^{2-} are shown in Figure 5.13. For O_2, there are two unpaired electrons in the antibonding π^*-MOs, meaning that O_2 is a diradical (oxidation state 0 for both atoms). The BO is 2, which is in agreement with the double bond character known for an O_2 molecule. For O_2^{2-}, both antibonding π^*-MOs are fully occupied with paired electrons, resulting in the oxidation state of −1 for both atoms. The two additional electrons come from both V(IV) centers, which are oxidized to V(V) again by transferring one electron each due to the π backdonation. For the O_2^{2-} ligand, the BO is 1, which is in agreement with the MO theory and the single bond between both atoms. Due to its two unpaired electrons, the O_2 molecule is paramagnetic. The O_2^{2-} ligand does not have any unpaired electrons, just like the

V(V) centers, so both the peroxo complex and the O_2^{2-} ligand are diamagnetic.[1,68,69]

5.4.2.2 Oxidation of Alkenes

Oxidation of alkenes yields epoxides, which hydrolyze to diols after contact with water. POM $[SiW_{10}O_{36}]^{8-}$, especially the anion $H_4[SiW_{10}O_{36}]^{4-}$, shows a very selective oxidation behavior for alkenes to epoxides with H_2O_2 at 305 K. For aliphatic C3–C8 alkenes, the epoxides are formed with 90% yield and a selectivity of >99%. No isomerization or cleavage of the C=C double bond was observed. The oxidant for the epoxidation is formed by the reaction of $H_4[SiW_{10}O_{36}]^{4-}$ with H_2O_2 and is called the hydroperoxide W–OOH. In the final step, one oxygen atom of the peroxido ligand is transferred to the alkene substrate (Figure 5.14).[12,70–72]

The hydroperoxido complex is another example of a peroxido complex (see Figure 5.13), where the peroxo ligand can coordinate to one or more metals in different motifs, as shown in Figure 5.14. Here, the peroxo ligand coordinates terminally to the metal and in Figure 5.13 the peroxo species bridges two metal centers in the POM cluster.

An oxygen atom of the peroxido ligand is transferred to the alkene, forming an epoxide. In contact with water, the epoxide is hydrolyzed to a diol species.

5.4.2.3 Bond Cleavages

The POM acid $H_5[PV_2Mo_{10}O_{40}]$ can insert an oxygen atom into the Sn–C bond (*e.g.* in tetrabutyltin, nBu$_4$Sn), yielding (nBu$_3$Sn)$_2$O and primary alcohols (*e.g.* 1-butanol, nBuOH) as products. This reaction works catalytically under aerobic conditions (under an O_2 atmosphere) and is not catalytic under anaerobic conditions (no O_2 atmosphere). Under anaerobic conditions, the olefines 1-butene, 2-butene and 2-methylpropene are also found as by-products. This reaction is typically done in acetonitrile.[12,73]

It is assumed that compounds like $(n\mathrm{Bu_4Sn})_x\mathrm{H}_{5+a-x}[\mathrm{PV}_a^{\mathrm{IV}}\mathrm{V}_{2-a}^{\mathrm{V}}\mathrm{Mo_{10}O_{40}}]$ are possible intermediates, which are formed by a one-electron transfer of nBu$_4$Sn to the POM (forming $n\mathrm{Bu_4Sn}^+$ cations), which reduces a V(V) species to V(IV). Under anaerobic conditions, the oxygen atoms in nBuOH and (nBu$_3$Sn)$_2$O come from the POM, yielding a reduced POM defect species of the type $\mathrm{H}_5\left[\mathrm{PV}_2^{\mathrm{IV}}\mathrm{Mo_{10}O_{39}}\right]$ as a by-product. This type of reaction is not catalytic anymore, because the POM gets consumed during the reaction. Under aerobic conditions, it can be shown by using $^{18}O_2$ as

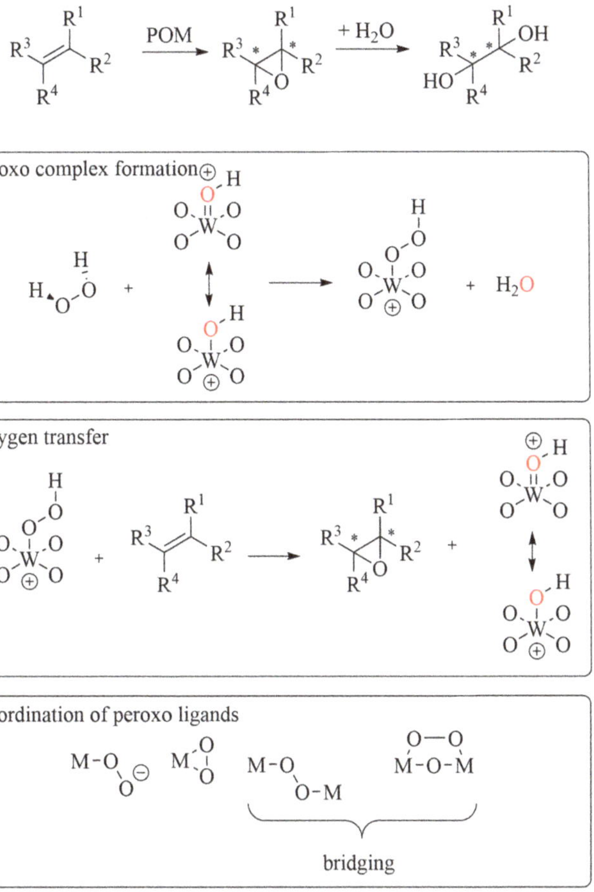

Figure 5.14 Mechanism for alkene epoxidation, catalyzed by POM $H_4[SiW_{10}O_{36}]^{4-}$.

an oxidant that the oxygen atom in *n*BuOH is from $^{18}O_2$ and the oxygen atom of $(n\text{Bu}_3\text{Sn})_2\text{O}$ comes from the POM. However, the POM is regenerated at the end, meaning that the reaction under aerobic conditions is a catalytic reaction. A summary of the reaction sequence is shown in the following equations under both conditions: oxygen insertion into the Sn–C bond under aerobic (catalytic) and anaerobic conditions (not catalytic).[12,73]

Aerobic conditions:

$$2n\text{Bu}_4\text{Sn} \xrightarrow[\text{POMH}_5[\text{PV}_2\text{Mo}_{10}\text{O}_{40}]]{+\text{O}_2,\ +\text{H}_2\text{O}} (n\text{Bu}_3\text{Sn})_2\text{O} + n\text{BuOH}$$

Anaerobic conditions:

$$2H_5[PV_2Mo_{10}O_{40}] + 2nBu_4Sn \rightarrow 2(nBu_4Sn)H_5[PV_2^{IV}Mo_{10}O_{40}]$$
$$\rightarrow (nBu_3Sn)_2O + nBuOH + C_4H_8$$
$$+ H_5[PV_2^{IV}Mo_{10}O_{39}]$$

The POMs $H_{3+x}[PV_xMo_{12-x}O_{40}]$ can also be used for the catalytic C–C bond cleavage of cyclohexane-1,2-diol, 2-hydroxycyclohexan-1-one and 2-methylcyclohexan-1-one, yielding adipic acid and 6-oxoheptanoic acid, as shown in Figure 5.15. Here, the VO_2^+ cation was identified as the catalytically active species. The mechanism is discussed using 2-hydroxycyclohexan-1-one as an example.[74]

- Step 1: 2-Hydroxycyclohexan-1-one is coordinated to the VO_2^+ cation (ligand association). The proton of the OH group on C2 migrates to the V=O oxo ligand of the VO_2^+ cation and forms a hydroxido ligand V–OH.[74]
- Step 2: The Lewis acidity of the V(v) species induces a keto-enol equilibrium (the proton of C2 migrates to the carbonyl oxygen atom C=O on C1). A double bond between C1 and C2 is formed.[74]
- Step 3: The V(v) center is reduced to V(iv) by accepting one electron of the V–OC2 bond. A radical electron remains on the oxygen atom, which recombines with one electron from the C1=C2 double bond at C2, forming a new C=O double bond between C2 and the oxygen atom. A new radical is formed on the carbonyl carbon atom on C1.[74]
- Step 4: Due to the radical character of molecular O_2 (Figure 5.13) it is able to recombine with the radical electron of V(iv) and the unpaired electron of C1, building a bridge between C1 and the V(iv) center. The C–O–O–V bridge can be interpreted as a peroxido ligand (compare Figures 5.13 and 5.14). Here, the electron is transferred from the V(iv) center to the antibonding π^*-MO of the CO_2^- peroxo ligand, resulting in an oxidation of V(iv) to V(v) (reoxidation step, compare Figure 5.13).[74]
- Step 5: The peroxido ligand is split, forming two radicals on the oxygen atoms. The C–C bond between C1 and C2 is also split, resulting in two radicals on C1 and C2. Now, the V(v) and C1 center each contain a single radical oxo ligand V(v)–O* and C1–O*. Two unpaired radical electrons also remain on C1 and C2 as a result of the C1–C2 single bond cleavage.[74]
- Step 6: The radical electron of C2 is transferred to the radical oxo ligand V(v)–O*, forming a V=O double bond between oxygen and

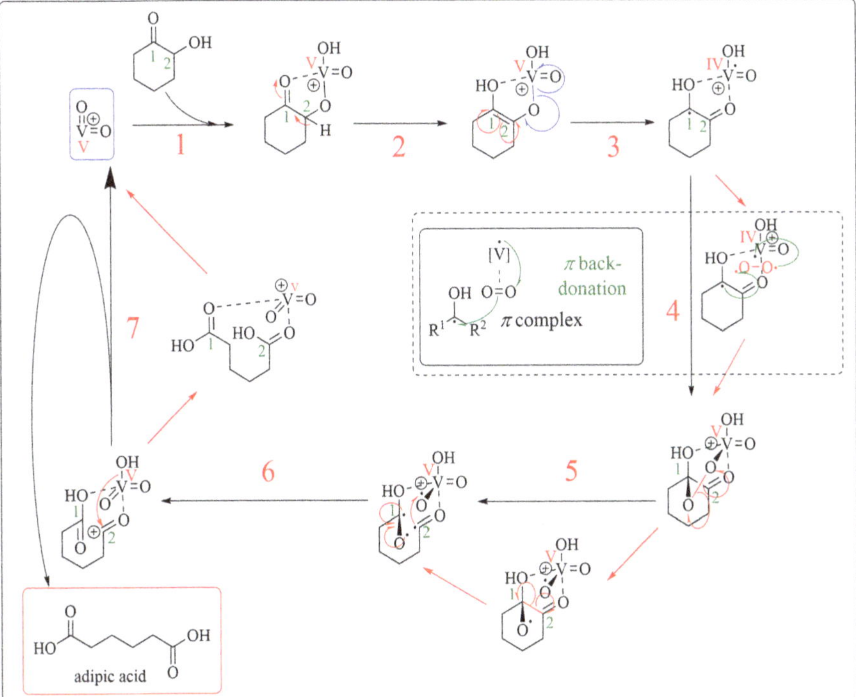

Figure 5.15 Catalytic C–C cleavage, catalyzed by V(v) containing POM acids. The catalytically active species is the VO_2^+ cation.[74]

the V(v) center. A cation remains at C2 as a result of the ensuing electron deficiency. This step is facilitated by the fact that oxygen is the more electronegative atom, drawing the electron toward itself. The cationic charge at the V(v) center disappears, as V(v) is now saturated with five covalent bonds: two V=O and one V–OH.

Both unpaired electrons on the radical oxygen (C1*–O*) atom and on C1 recombine, forming a new carboxylate species on C1.[74]
- Step 7: The hydroxido ligand on V(v) (V–OH) recombines with the cation on C2, yielding a second carboxylate species on C2. Adipic acid is formed, which can still coordinate to the VO_2^+ cation. In the last step, adipic acid dissociates from V(v), regenerating the fully intact VO_2^+ catalyst species.[74]

5.4.3 Reduction Catalysis

In general, the use of POM catalysts in reduction catalysis is less explored compared to oxidative catalysis. Well-known examples of the use of POMs in reductive catalytic applications are the reduction of aldehydes and the hydrogenation of aromatic ketones.[12]

5.4.3.1 *Homogeneous and Heterogeneous Catalyzed Reductions*

For the reduction of aldehydes, molecular H_2 is activated on a square planar Pt complex. With the activated H_2 the V(v) in the POM acid $H_5[PV_2VMo_{10}O_{40}]$ is reduced to V(iv), forming the reduced POM acid $H_7[PV_2^{IV}Mo_{10}O_{40}]$, which can transfer two electrons to an aldehyde (*e.g.* benzaldehyde). V(iv) will be reoxidized to V(v), forming a cationic species $H_7[PV_2^{V}Mo_{10}O_{40}]^{2+}$. The aldehyde forms a dianionic species, which couples directly with another aldehyde, forming an anionic vicinal diol. This anionic species can abstract two protons from the cationic species $H_7[PV_2^{V}Mo_{10}O_{40}]^{2+}$, yielding the vicinal diol as a product under regeneration of the catalyst $H_5[PV_2^{V}Mo_{10}O_{40}]$. A vicinal diol tends to rearrange under acidic conditions. One of the hydroxyl groups is protonated and eliminated as a water molecule. A carbenium cation is formed. The phenyl group can now migrate in a [1,2] shift, forming a more stable cation, yielding an aldehyde after deprotonation. This rearrangement of the vicinal diol to the aldehyde is called the Pinacol rearrangement. A possible catalytic cycle and a mechanism for the rearrangement are shown in Figure 5.16.[75]

The second reductive catalytic application for POMs is a heterogeneous catalyzed reaction with a Pd(ii) substituted POM $K_5[PPdW_{11}O_{39}]$, which is supported on γ-alumina (Al_2O_3) or active carbon ($K_5[PPdW_{11}O_{39}]/Al_2O_3$ or $K_5[PPdW_{11}O_{39}]/C$). The preparation of these supported catalysts is an easy method to heterogenize homogeneous catalysts and to benefit from the advantages of a heterogeneous catalyst system. These catalyst

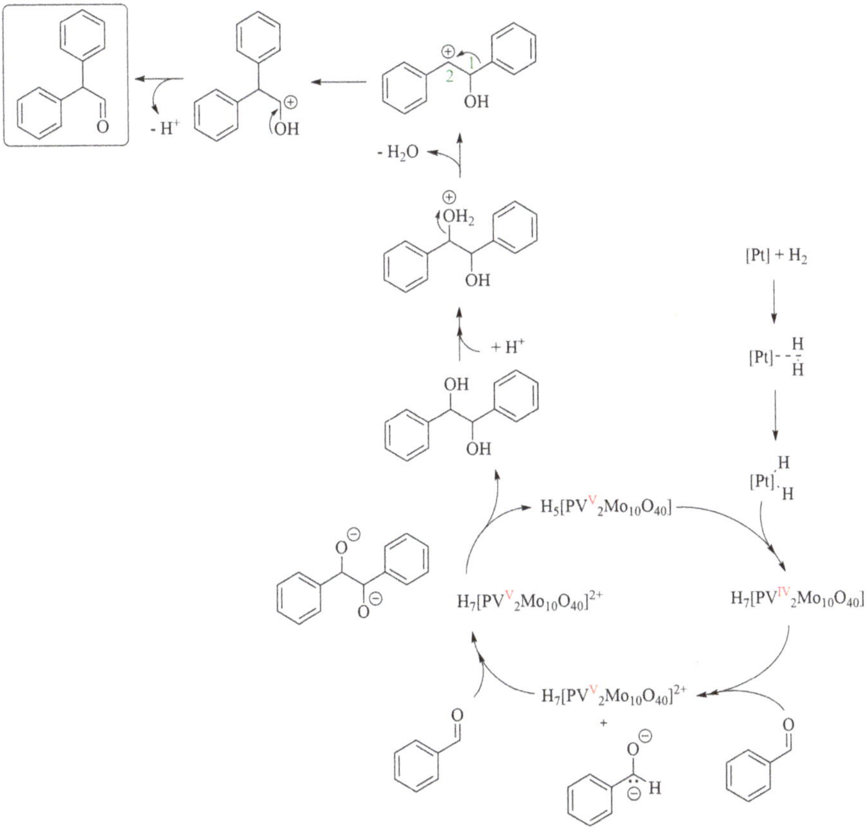

Figure 5.16 Catalytic reduction of aldehydes and subsequent Pinacol rearrangement with a POM acid. Molecular H_2 is activated by a Pt complex, which reduces the POM. In the next step the POM can reduce the substrate. Vicinal diols can react in a Pinacol rearrangement in acidic media.

systems are very active for the hydrogenation of aromatic compounds like toluene to methylcyclohexane, as shown in Figure 5.17.[76]

A noticeable trend can be observed for aromatic ketone compounds, for which a selective hydrogenation of the aromatic nucleus is possible. Simple aromatic ketones, such as benzophenone or acetophenone, are completely hydrogenated (on the ketone functionality and on the aromatic nucleus), while ketones like 1-phenylpropan-2-one and 1-phenylbutan-2-one were hydrogenated to 1-cyclohexylpropan-2-one and 1-cyclohexylbutan-2-one. This observation differs from the commercial Pd catalyst on carbon (Pd/C), which hydrogenates the aromatic ring and the ketone to the corresponding alcohol. Aromatic ketones are directly deoxygenated to the methylene moiety by Pd/C.[76]

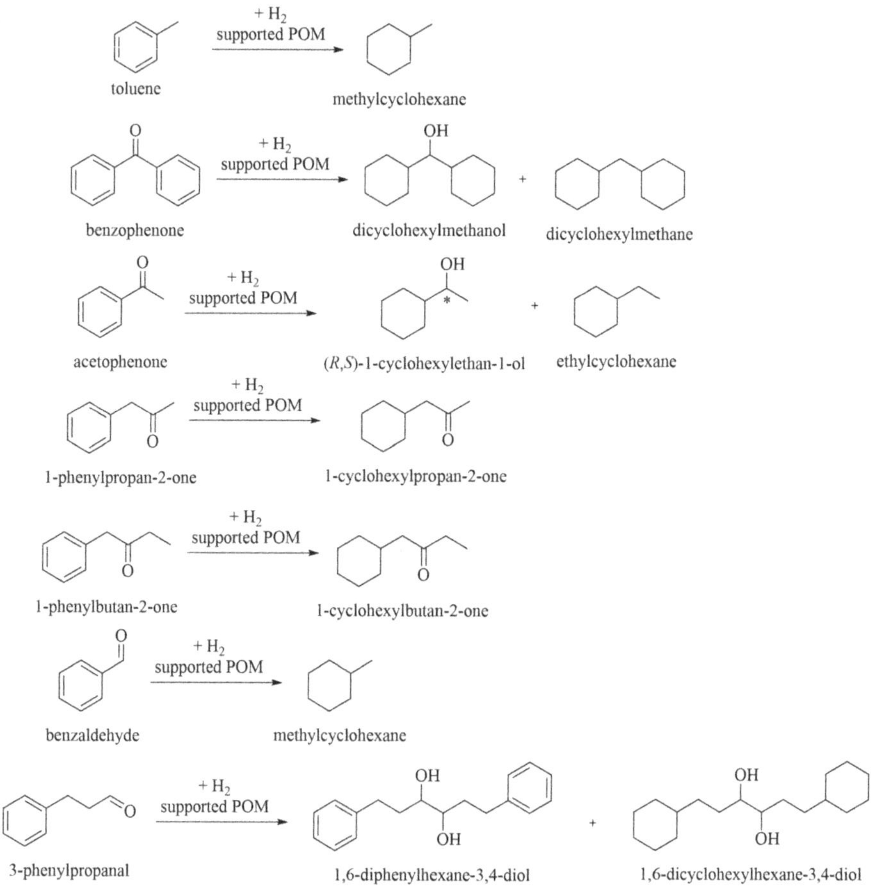

Figure 5.17 Catalytic hydrogenation of aromatic ketones and aldehydes by supported POM catalysts.

Aldehydes like benzaldehyde are directly reduced to methylcyclohexane, while aldehydes like 3-phenylpropanal are converted to a dimeric diol compound, followed by a hydrogenation of the aromatic rings.[76]

Another well-known reductive application is homogeneous hydroformylation, HyFo (also known as the oxo synthesis or Roelen reaction), a very prominent chemical reaction used to produce aldehydes from alkenes and syngas (H_2 and CO). Industrially, HyFo is done using Co(I) or Rh(I) catalysts. Since Rh(I) is expensive, Co(I)-based catalysts are typically studied in laboratory experiments, such as the famous catalyst tetracarbonylhydridocobalt(I) [$CoH(CO)_4$]. Aldehydes are of fundamental economic importance because the carbon atom in the aldehyde functional group is in the +I oxidation state. This makes aldehydes ideal starting materials for subsequent reductions (to alcohols with an oxidation state of −I) or oxidations (to carboxylic acids with an oxidation state of +III). The rich downstream chemistry of aldehydes (including the aldol reaction) makes them fundamental precursors for further products. For example, aldehydes can be used to synthesize pharmaceuticals such as ibuprofen.[1,2,7]

It has been shown that Co(II)-substituted POMs of the phosphomolybdate type $H_{3+4x}[PCo_xMo_{12-x}O_{40}]$ can also be used as catalysts in HyFo. The oxo synthesis of 1-hexene to the aldehydes *n*- and iso-heptanal was studied in a thermomorphic solvent system (Figure 5.3) consisting of water (POM phase) and 1-butanol (1-hexene and aldehyde phase). Since the anionic charge increases by 4 units with each additional substitution in the POM systems, the resulting anionic species are: −7 for $x=1$, −11 for $x=2$, and −15 for $x=3$. It was shown that $x=3$ is the highest possible degree of substitution before the anionic charge becomes too high and the TMSPOM can no longer be stabilized.[7]

Typical reaction conditions for HyFo are 130 °C, 150 bar of syngas pressure, a $CO:H_2$ ratio of 1:1, and a reaction time of 4 hours. For $x=2$, the *n*- and iso-heptanal yields are 14% and 5%, with a 1-hexene conversion of 64%. For $x=1$ and the pure phosphomolybdate, the *n*- and iso-heptanal yields are 3% and 2%, with a total 1-hexene conversion of 36%. TMSPOM with $x=3$ does not show sufficient stability under process conditions. Thus, HyFo activity increases steadily with increasing degrees of substitution.[7]

It was observed that at a $CO:H_2$ gas ratio of 2:1, the lowest conversions and the lowest aldehyde yields were obtained. At a gas ratio of 1:2, the conversion was nearly complete, and the yield of *n*-heptanal reached 46%. An approximate *n*:iso ratio of 3:1 was observed. Under 130 °C, HyFo came to a standstill. As the temperature increased in the range of 130 to 140 °C, the conversion rose to nearly 100%, and the aldehyde product yields increased steadily. Here as well, an *n*:iso ratio of 3:1 was observed. The optimal reaction conditions were determined to be a $CO:H_2$ ratio of 1:2, a total gas pressure of 150 bar, a reaction temperature of 130 °C, and a reaction time of four hours.

Figure 5.18 shows the typical catalytic cycle of HyFo involving a Co(I) complex, featuring the classical reaction steps: olefin association, olefin insertion, CO association, CO insertion, oxidative addition of H_2, and reductive elimination of the aldehyde. For the POMs, a dissociative catalytic mechanism in aqueous medium has been

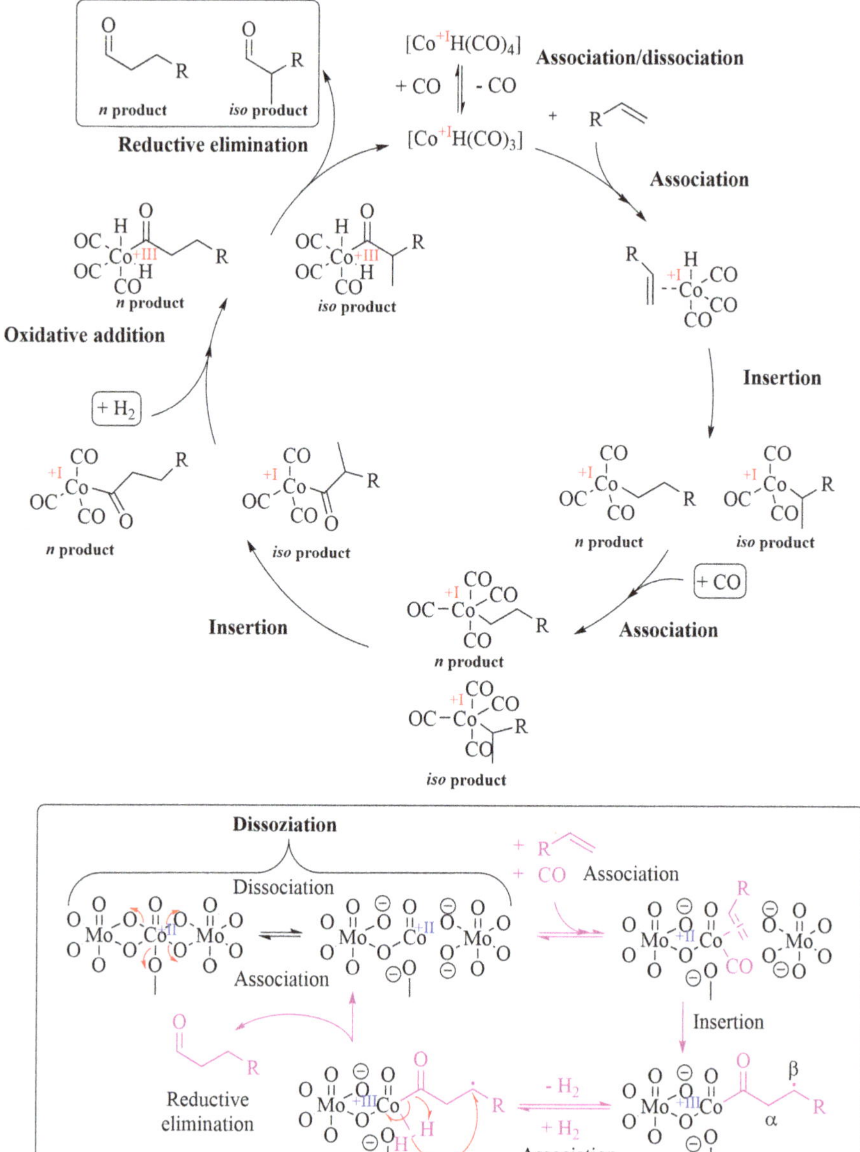

Figure 5.18 Mechanism of Co(I)-catalyzed HyFo with its typical steps (top): (olefin) association, (olefin) insertion, (CO) association, (CO) insertion, oxidative addition of H_2, and reductive elimination of the aldehyde. Postulated reaction mechanism for HyFo with Co(II)-substituted POMs (bottom).

proposed, in which a Co(II) oxo species partially dissociates from the POM framework to create free coordination sites at the Co(II) center. The catalytic cycle could begin with the association of a CO ligand and the olefin. Via a radical mechanism, the olefin is inserted, leaving a radical electron on the β-carbon atom. In the next step, an H_2 ligand could coordinate to the Co(II) center, allowing for the reductive elimination of the aldehyde, with the β-carbon atom recombining with an additional hydrogen atom. The Co(II) center temporarily switches to the +III oxidation state during this process. It is also possible that this mechanism proceeds in a concerted manner (with bond breaking and formation occurring simultaneously), thereby avoiding the formation of radical intermediates.[1,2,7]

5.4.3.2 Photocatalytic Reductions

POMs can also take part in photocatalytic reductions. CO_2 is a frequently used C_1 renewable feedstock for preparing different chemicals. The development of a catalytic application that uses CO_2 as a substrate is considered green and sustainable chemistry. However, the interaction between the negatively charged oxo surface of the POM and CO_2 is not favored.

To solve this problem, special elements on the POM surface, like Co, Ni, Mn or Ru, can coordinate weakly bonded ligands (such as H_2O), which can be substituted by CO_2 in anhydrous organic solvents. Using a sacrificial electron donor, the bonded CO_2 can be reductively activated by irradiation with light. A mono-substituted Keggin-type POM $[SiRuW_{11}O_{39}]^{5-}$ was used, which can coordinate a water molecule on the Ru(III) center $[Ru(H_2O)SiW_{11}O_{39}]^{5-}$. In the next step, the water molecule is substituted by CO_2 yielding $[Ru(CO_2)SiW_{11}O_{39}]^{5-}$. Triethylamine can be used as a sacrificial electron donor and the CO_2 can be reduced under ultraviolet (UV) light irradiation in toluene solution, yielding CO as the main product. A catalytic transformation of CO_2 to CO is a valuable process, because CO can be used as a syngas equivalent in further catalytic applications. The catalytic mechanism is shown in Figure 5.19.[12,77]

- Water dissociates forming a RuO_5 species, which has a free coordination site for CO_2.[12,77]
- CO_2 coordinates in an end-on motif, forming a Ru–OCO species (ligand association). Triethylamine also coordinates to the complex, forming attractive Et_3N–CO_2 interactions. A C–H group of triethylamine forms a hydrogen bond interaction with a bridging oxo ligand bonded to the Ru(III) center, yielding a N–C–H–O interaction.[12,77]

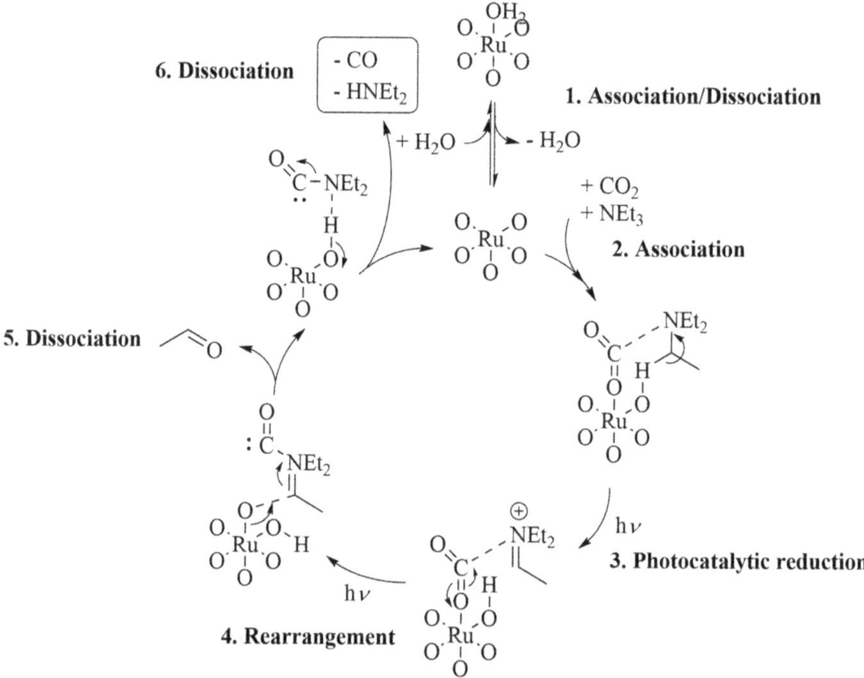

Figure 5.19 Possible catalytic mechanism for the photoreduction of CO_2 catalyzed by $[SiRuW_{11}O_{39}]^{5-}$.

- Facilitated by hydrogen bonding, a proton from the CH_2 moiety of triethylamine is transferred to the bridging oxo ligand coordinated to the Ru(III) center of the POM. A double bond between carbon and nitrogen is formed with a cationic charge on nitrogen: C=N. Cleavage of the N–C bond within the Et_2N–CO_2 moiety occurs, with the electron pair shifting onto the carbon atom of the CO_2 group. This leads to the formation of a carbenoid species and can be interpretated as a two electron reduction.[12,77]
- In the subsequent steps, the C–O bond in the CO_2 derived C–O–Ru motif is cleaved photochemically. An oxo ligand remains coordinated to the Ru(III) center, carrying a negative charge (Ru–O–). A carbonylcarbene species is formed, which engages in a non-covalent interaction with the nitrogen atom of the C=NEt_2 moiety.[12,77]
- One of the lone pairs on the negatively charged oxo ligand forms a covalent bond with the carbon atom bearing a positive formal charge. This leads to the formation of a Ru–O–C–N species. The lone pair on the nitrogen atom is protonated by a proton.

Simultaneously, the Ru–O bond and the C–N bond within the Ru–O–C–N framework are cleaved. Acetaldehyde is formed, along with a new covalent bond between the carbon atom of the carbonylcarbene and the nitrogen atom. The resulting species is an O–C–$NHEt_2$ moiety, bearing a negative charge on the oxygen atom and a positive formal charge on the nitrogen atom.[12,77]

- CO and diethylamine are formed, regenerating the catalytically active POM species with an open RuO_5 site on which a water molecule can coordinate.[12,77]

5.5 Photocatalytic Applications

The decatungstate $[W_{10}O_{32}]^{4-}$ anion is known to be reduced by irradiation with light in the presence of organic compounds or impurities. Dissolving the salt $TBA_4[W_{10}O_{32}]$ in *N*,*N*-dimethylformamide (DMF) in the presence of ethanol or poly(vinyl alcohol) leads to the formation of a blue solution, indicating the reduction of $[W_{10}O_{32}]^{4-}([W^{V}_{10}O_{32}]^{4-})$ to $[W_{10}O_{32}]^{5-}\left(\left[W^{IV}W^{V}_{9}O_{32}\right]^{5-}\right)$ (A) and/or the formation of anions of the type $H_x\left[W_{10}O_{32}\right]^{(6-x)-}\left(H_x\left[W^{IV}_{x}W^{V}_{10-x}O_{32}\right]^{(6-x)}\right)$ (B). (A) is the product of a one-electron reduction and (B) is the product of a two- or multi-electron reduction. With increasing irradiation time, the amount of species (B) increases. Only (B) is obtained after irradiation of more than one hour or in solutions with high ethanol concentrations.[78]

Due to its photocatalytic activity, $[W_{10}O_{32}]^{4-}$ has many applications in organic chemistry. A very well-known example is the acylation of electron-poor olefins. As shown in Figure 5.20, the catalytic cycle starts with a C–H bond cleavage of the aldehyde, forming an acyl radical. $\left[W^{VI}_{10}O_{32}\right]^{4-}$ is reduced to $\left[W^{V}W^{VI}_{9}O_{32}\right]^{5-}$ and the aldehyde proton is abstracted by the reduced anion. The acyl radical gets trapped by the electrophilic olefin, forming a secondary stabilized radical. Anion $\left[W^{V}W^{VI}_{9}O_{32}\right]^{5-}$ is reoxidized to the catalytically active species $\left[W^{VI}_{10}O_{32}\right]^{4-}$ and the electron is transferred to the secondary radical, forming a new C–H bond with the single proton. This reaction yields unsymmetrical ketones as the main product. The only limiting step of this reaction is the decarbonylation under CO release. Both reactions compete with each other, with decarbonylation occurring if tertiary or secondary aldehydes are used as substrates. The driving force behind decarbonylation is the formation of more stable tertiary radicals. At temperatures below −20 °C, decarbonylation is suppressed for secondary aldehydes and the formation of the acylation product is favored.[79–81]

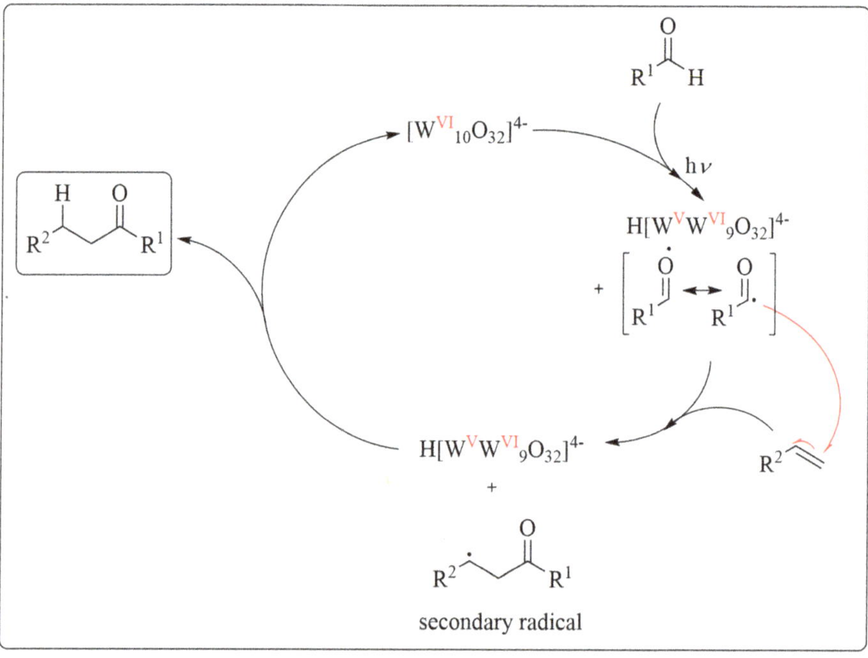

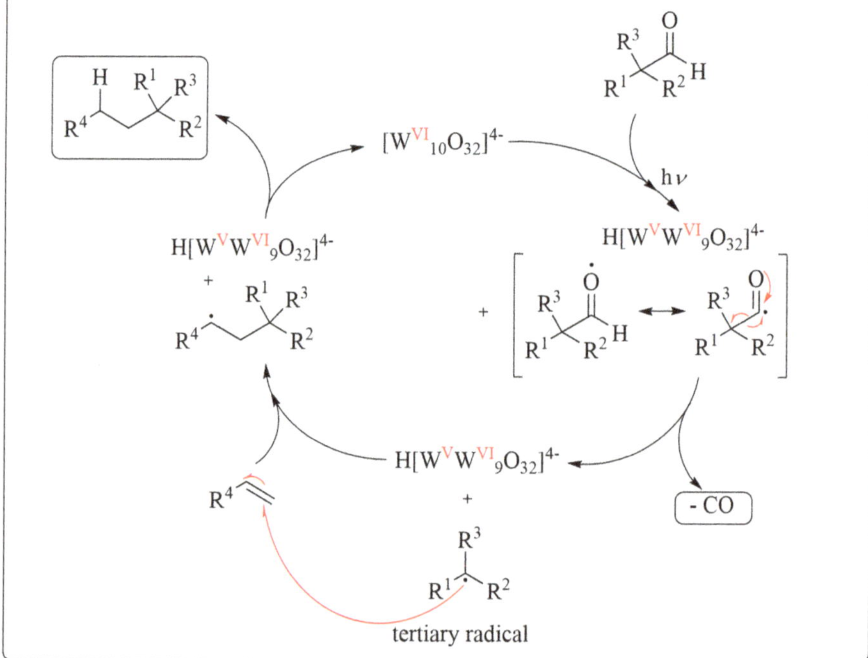

Figure 5.20 Acylation of electron poor olefins catalyzed by $[W_{10}O_{32}]^{4-}$. With secondary or tertiary aldehydes, a decarbonylation takes place as side reaction.

5.6 Cyclization Reactions

Different types of cyclization reactions are catalyzed by POMs, as shown in Figure 5.21.

The Cu(I) catalyzed cyclization of organic azides with alkynes is a very easy and efficient method for synthesizing aromatic triazol derivatives and is also known as a "click reaction". This is a 1,3-dipolar cycloaddition in which a 1,3-dipole (the azide) reacts with a dipolarophile (the alkyne). A 1,4-disubstituted 1,2,3-triazole derivative is formed as a reaction product, which is why this reaction is also referred to as copper catalyzed azide–alkyne cycloaddition (CuAAC). The cyclization takes place in a single, concerted reaction step.[82] POM $TBA_4[H_2SiCu_2(\mu\text{-}1,1\text{-}N_3)_2W_{10}O_{36}]$ is an efficient catalyst for this type of reaction. The 1,4-disubstituted 1,2,3-triazole derivative is formed with a high yield in acetonitrile as a solvent.[83] A typical mechanism for the Cu(I) cation catalyzed 1,3-dipolar cycloaddition is shown in Figure 5.22, which is done with CuI only.[82]

- The alkyne is coordinated to the Cu(I) center, increasing the acidity of the alkyne proton.[82]
- The proton is split off, forming a C–Cu bond. This species is called an acetylide, in which the second carbon atom of the alkyne is now positively polarized and activated for a nucleophilic attack of the terminal nitrogen atom of the azide.[82]

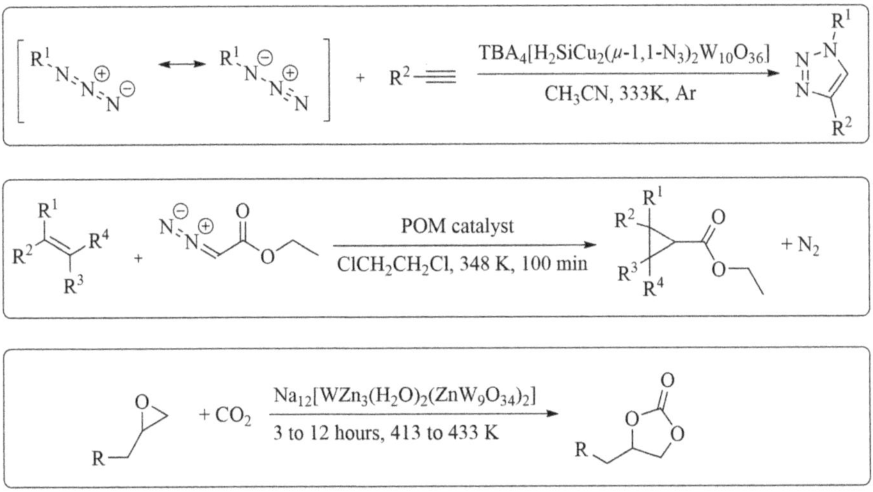

$TBA_4[H_2SiCu_2(\mu\text{-}1,1\text{-}N_3)_2W_{10}O_{36}]$, CH_3CN, 333K, Ar

POM catalyst, $ClCH_2CH_2Cl$, 348 K, 100 min; $+ N_2$

$+ CO_2$, $Na_{12}[WZn_3(H_2O)_2(ZnW_9O_{34})_2]$, 3 to 12 hours, 413 to 433 K

Figure 5.21 Overview of possible cyclization reactions in which POMs are involved as catalysts.

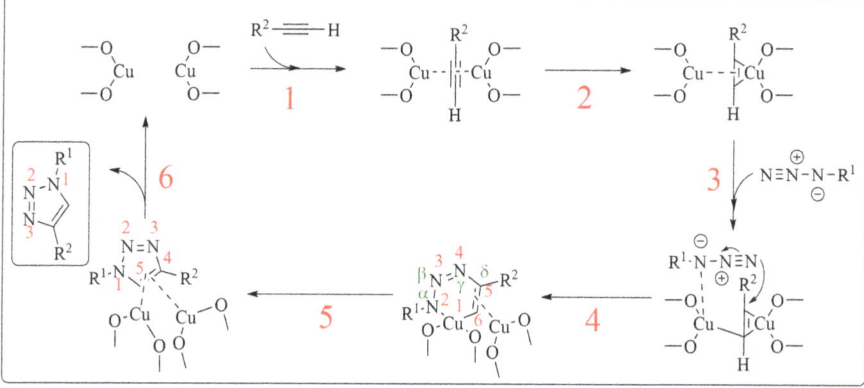

Figure 5.22 Reaction mechanisms for the Cu(I) catalyzed 1,3-dipolar cycloaddition (top) and the Cu(I) POM catalyzed cycloaddition (bottom).

- The azide is coordinated to the Cu(I) center.[82]
- The cyclization takes place by a nucleophilic attack of the terminal nitrogen atom on the positively polarized carbon atom, forming a six-membered ring containing the Cu(I) center. This species is called an α,δ-disubstituted 2,3,4-triazole or dicopper(I) metallacycle.[82]
- The 1,4-disubstituted 1,2,3-triazol derivative is formed by forming a bond between N2 and C6, which is still coordinated to the Cu(I) center with C5.[82]

- The 1,4-disubstituted 1,2,3-triazol is formed and the catalyst species Cu(I) is regenerated.[82]

The POM catalyzed reaction requires both Cu(I) centers, as shown in Figure 5.22.[83]

- The alkyne is coordinated to both Cu(I) centers.[83]
- The coordinated alkyne forms a metallacyclopropene ring.[83]
- The azide starts to interact with the metallacyclopropene, forming a monoalkynyl monoazido dicopper(I) intermediate.[83]
- The terminal nitrogen atom of the azide starts a nucleophilic attack on the positively charged carbon atom of the acetylide species. A six-membered ring is formed, called a dicopper(I) metallacycle (α,δ-disubstituted 2,3,4-triazole).[83]
- A bond between N2 and C6 is formed, yielding the 1,4-disubstituted 1,2,3-triazol derivative, which is coordinated to both Cu(I) centers.[83]
- The 1,4-disubstituted 1,2,3-triazol is released and the active catalyst species is regenerated.[83]

Both mechanisms are very similar, but instead of forming a C–Cu species, a metallacyclopropene species is formed in the POM catalyzed case. The difference is that the alkyne does not lose its terminal proton, which means that the dicopper(I) metallacycle and 1,4-disubstituted 1,2,3-triazol species can only coordinate to the Cu(I) center *via* their π electrons.

POM anions $[PW_{11}O_{39}Co(H_2O)]^{5-}$, $[PW_{11}O_{39}Cu(H_2O)]^{5-}$ and the Krebs-type anion $[Co_2(H_2O)_6\{W(OH)_2\}_2(BiW_9O_{33})_2]^{6-}$ can catalyze cyclopropanation reactions, as shown in Figure 5.21. For this type of reaction, only the catalyst, the cyclopropane derivative and ethyl diazoacetate (EDA) is required. Reactions are performed in 1,2-dichloroethane as solvent. A driving force of this reaction is the release of molecular N_2 gas.[84]

Another POM-catalyzed reaction is the cycloaddition of CO_2 to epoxides. This reaction is catalyzed by a sandwich-type POM $Na_{12}[WZn_3(H_2O)_2(ZnW_9O_{34})_2]\cdot 46H_2O$, which is a dimer of two Keggin lacunary-type structures, containing Zn(II) as heteroelement. Both lacunary-type units are coordinated to a W(VI) and three Zn(II) cations (see Chapter 4, Section 4.1.1). Typical experimental procedures are performed with the epoxides, CO_2, the catalyst and dimethyl amino pyridine (DMAP) in dichloromethane (DCM).

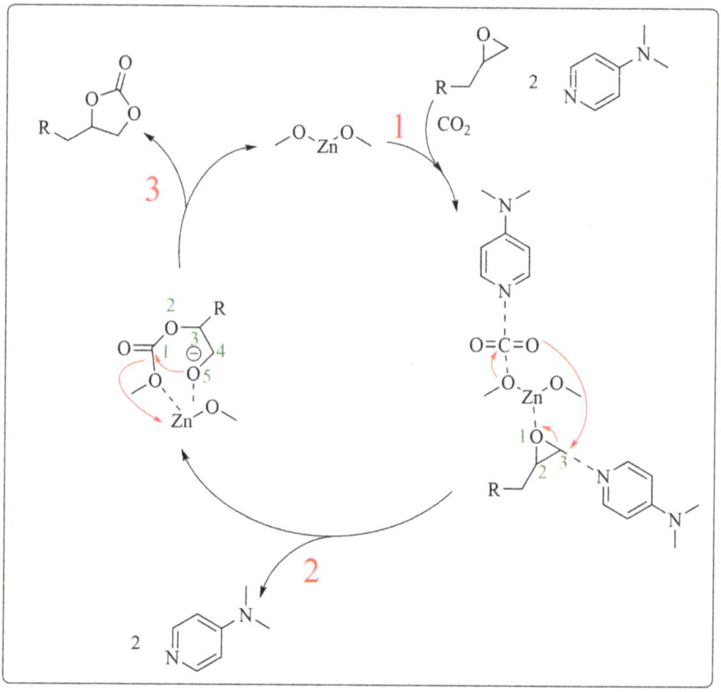

Figure 5.23 Catalytic cycle of the cycloaddition of CO_2 to an epoxide.

DMAP is used as a Lewis base. This reaction is a very effective way to use CO_2 that is formed in different reactions as an unwanted by-product. The catalytic mechanism is shown in Figure 5.23.[85]

- In the first step, the oxygen atom of the epoxide is coordinated to the Zn(II) center. CO_2 is coordinated to the neighboring oxo ligand of Zn(II). The catalytic effect is supported by a co-operative effect between DMAP and the Zn-POM. Here, the epoxide is activated by the interaction with the Lewis acidic Zn(II) center and by a nucleophilic interaction between DMAP and the C3 of the epoxide. A comparable effect is observed for CO_2: the carbon atom of CO_2 is positively polarized by coordinating to the oxo ligand and by a nucleophilic interaction of DMAP.[85]
- In the second step, a nucleophilic attack of the oxygen atom of CO_2 takes place on C3. The oxo ligand is split off from Zn(II) and binds to the carbon atom of CO_2. However, a coordinative bond between the oxo ligand and Zn(II) remains. An oxo anion is formed from the epoxide O1 atom, which coordinates to Zn(II).[85]

- In the third step, a nucleophilic attack of O5 with C1 takes place, forming the five-membered ring. The oxo ligand coordinates to Zn(II) again, regenerating the fully intact catalytically active species and the reaction product.[85]

Note: the catalytic reaction takes place on the Zn(II) cations complexed by the two lacunary-type POMs, not on the heteroelement, which is much more difficult to access.

Abbreviations

POM	Polyoxometalate
RedOx	Reduction/oxidation
FA	Formic acid
IR	Infrared
UV	Ultraviolet
LA	Levulinic acid
HMF	Hydroxymethylfurfural
DES	Deep eutectic solvents
HBA	Hydrogen bond acceptor
HBD	Hydrogen bond donor
DHH	6-Hydroxy-2,5-dioxohexanal
TMSCN	Trimethylsilyl cyanide
OxFA	Oxidative formation of formic acid
OA	Oxalic acid
pTSA	*para*-Toluene sulfonic acid
IL	Ionic liquid
AA	Acetic acid
ECODS	Extraction-coupled oxidative desulfurization
ODS	Oxidative desulfurization
SAA	Sulfoacetic acid
SAA	Sulfoacetic acid
2-SBA	2-Sulfobenzoic acid
2-SOBA	2-(Sulfooxy)benzoic acid
MO	Molecular orbital
BO	Bond order
HyFo	Hydroformylation
UV	Ultraviolet
DMF	*N*,*N*-Dimethylformamide
CuAAC	Copper catalyzed azide–alkyne cycloaddition
EDA	Ethyldiazoacetate

DMAP	Dimethyl amino pyridine
DCM	Dichloromethane

Acknowledgements

I would like to thank the publisher, the Royal Society of Chemistry, for the opportunity to write this book!

Recommended Reading

Please have a look into the following reference:

1. J.-C. Raabe, M. J. Poller, D. Voß and J. Albert, *ChemSusChem*, 2023, **16**, e202300072.

References

1. D. Steinborn, in *Grundlagen Der Metallorganischen Komplexkatalyse*, Springer, Berlin, Heidelberg, 2019.
2. A. Behr, D. W. Agar, J. Jörissen and A. J. Vorholt, in *Einführung in Die Technische Chemie*, Springer, Berlin, Heidelberg, 2016.
3. A. K. Singh, S. Singh and A. Kumar, *Catal. Sci. Technol.*, 2016, **6**, 12–40.
4. S. Enthaler, J. von Langermann and T. Schmidt, *Energy Environ. Sci.*, 2010, **3**, 1207–1217.
5. J. M. Berg, J. L. Tymoczko, G. J. Gatto and L. Stryer, in *Stryer Biochemie*, Springer, Berlin, Heidelberg, 2018.
6. J.-C. Raabe, M. J. Poller, D. Voß and J. Albert, *ChemSusChem*, 2023, **16**, e202300072.
7. J.-C. Raabe, L. Hombach, M. J. Poller, A. Collauto, M. M. Roessler, A. Vorholt, A. K. Beine and J. Albert, *ChemCatChem*, 2024, **16**, e202400395.
8. I. V. Kozhevnikov, *Russ. Chem. Rev.*, 1987, **56**, 811–825.
9. I. V. Kozhevnikov, *Chem. Rev.*, 1998, **98**, 171–198.
10. J. Reichert, B. Brunner, A. Jess, P. Wasserscheid and J. Albert, *Energy Environ. Sci.*, 2015, **8**, 2985–2990.
11. S. Maerten, C. Kumpidet, D. Voß, A. Bukowski, P. Wasserscheid and J. Albert, *Green Chem.*, 2020, **22**, 4311–4320.
12. S.-S. Wang and G.-Y. Yang, *Chem. Rev.*, 2015, **115**, 4893–4962.
13. S. Wesinger, M. Mendt and J. Albert, *ChemCatChem*, 2021, **13**, 3662–3670.
14. T. Esser, A. Wassenberg, D. Voß and J. Albert, *React. Chem. Eng.*, 2024, **9**, 1666–1684.
15. A. Wassenberg, T. Esser, M. J. Poller and J. Albert, *Materials*, 2023, **16**, 2864.
16. A. Wassenberg, T. Esser, M. J. Poller, D. Voß and J. Albert, *Biofuels, Bioprod. Biorefin.*, 2024, **18**, 1585–1597.
17. T. Esser, A. Wassenberg, D. Voß and J. Albert, *Chem. Eng. Res. Des.*, 2024, **209**, 311–322.
18. T. Esser, A. Wassenberg, J. Raabe, D. Voß and J. Albert, *ACS Sustainable Chem. Eng.*, 2024, **12**, 543–560.
19. S. G. Maerten, D. Voß, M. A. Liauw and J. Albert, *ChemistrySelect*, 2017, **1**, 1–8.

20. S. Körner, J. Albert and C. Held, *Front. Chem.*, 2019, 7, 1–11.
21. J. Macht, M. J. Janik, M. Neurock and E. Iglesia, *Angew. Chem., Int. Ed.*, 2007, **46**, 7864–7868.
22. Z.-Z. Weng, J. Xie, K.-X. Huang, J.-P. Li, L.-S. Long, X.-J. Kong and L.-S. Zheng, *Inorg. Chem.*, 2022, **61**, 4121–4129.
23. A. R. Gaspar, J. A. F. Gamelas, D. V. Evtuguin and C. P. Neto, *Green Chem.*, 2007, **9**, 717–773.
24. A. A. Shatalov, D. V. Evtuguin and C. P. Neto, *Carbohydr. Polym.*, 2000, **43**, 23–32.
25. I. A. Weinstock, R. H. Atalla, U. P. Agarwal, J. L. Minor and C. Petty, *Spectrochim. Acta, Part A*, 1993, **49**, 819–829.
26. D. V. Evtuguin, C. Pascoal Neto, J. Rocha and J. D. Pedrosa de Jesus, *Appl. Catal., A*, 1998, **167**, 123–139.
27. M. Suchy and D. S. Argyropoulos, in *Oxidative Delignification Chemistry*, UTC, 2001, pp. 2–43.
28. D. V. Evtuguin, A. I. D. Daniel, A. J. D. Silvestre, F. M. L. Amado and C. Pascoal Neto, *J. Mol. Catal. A: Chem.*, 2000, **154**, 217–224.
29. A. Gaspar, D. V. Evtuguin and C. P. Neto, *Appl. Catal., A*, 2003, **239**, 157–168.
30. H. Lange, S. Decina and C. Crestini, *Eur. Polym. J.*, 2013, 1151–1173.
31. A. R. Mankar, A. Modak and K. K. Pant, *Adv. Sustainable Syst.*, 2021, 1–25.
32. D. V. Evtuguin, C. P. Neto, H. M. Carapuça and J. Soares, *Holzforschung*, 2000, **54**, 511–518.
33. R. Ma, M. Guo and X. Zhang, *Catal. Today*, 2018, **302**, 50–60.
34. A. A. Ibrahim, M. El-Sakhawy and S. Kamel, *Polpu, Chongi Gisul/J. Korea Tech. Assoc. Pulp Pap. Ind.*, 2005, **37**, 56–62.
35. A. A. Shatalov and H. Pereira, in *6° Congresso Florestal Nacional*, Ponta Delgada, Port, 2009, pp. 572–577.
36. A. A. Shatalov and H. Pereira, *Bioresour. Technol.*, 2010, **101**, 9330–9334.
37. A. A. Shatalov and H. Pereira, *Bioresour. Technol.*, 2010, **101**, 4616–4621.
38. A. A. Shatalov, *Green Chem.*, 2017, **19**, 5092–5102.
39. V. Ahuja, in *Analysis of Novel Transition Metals as Catalysts for Oxygen Delignification*, 2001.
40. I. A. Weinstock, R. E. Schreiber and R. Neumann, *Chem. Rev.*, 2018, **118**, 2680–2717.
41. C. K. Tiwari, M. Baranov, A. Neyman, R. Neumann and I. A. Weinstock, *Inorg. Chem.*, 2020, **59**, 11945–11952.
42. P. Preuster and J. Albert, *Energy Technol.*, 2018, **6**, 501–509.
43. J. Albert, R. Wölfel, A. Bösmann and P. Wasserscheid, *Energy Environ. Sci.*, 2012, **5**, 7956–7962.
44. J. Albert, D. Lüders, A. Bösmann, D. M. Guldi and P. Wasserscheid, *Green Chem.*, 2014, **16**, 226–237.
45. R. Wölfel, N. Taccardi, A. Bösmann and P. Wasserscheid, *Green Chem.*, 2011, **13**, 2759–2763.
46. S. Ponce, M. Trabold, A. Drochner, J. Albert and B. J. M. Etzold, *Chem. Eng. J.*, 2019, **369**, 443–450.
47. Z. Zhang and G. W. Huber, *Chem. Soc. Rev.*, 2018, **47**, 1351–1390.
48. D. Voß, M. Kahl and J. Albert, *ACS Sustainable Chem. Eng.*, 2020, **8**, 10444–10453.
49. J. Albert and P. Wasserscheid, *Green Chem.*, 2015, **17**, 5164–5171.
50. S. Ponce, S. Wesinger, D. Ona, D. A. Streitwieser and J. Albert, *Biomass Convers. Biorefin.*, 2023, **13**, 7199–7206.
51. J. Albert, *Faraday Discuss.*, 2017, **202**, 99–109.
52. N. V. Gromov, T. B. Medvedeva, K. N. Sorokina, Y. V. Samoylova, Y. A. Rodikova and V. N. Parmon, *ACS Sustainable Chem. Eng.*, 2020, **8**, 18947–18956.
53. J. Reichert and J. Albert, *ACS Sustainable Chem. Eng.*, 2017, **5**, 7383–7392.
54. J. Albert, M. Mendt, M. Mozer and D. Voß, *Appl. Catal., A*, 2019, **570**, 262–270.

55. Y. Hou, M. Niu and W. Wu, *Ind. Eng. Chem. Res.*, 2020, **59**, 16899–16910.
56. S. Wesinger, M. Mendt and J. Albert, *ChemCatChem*, 2021, **13**, 3662–3670.
57. A. Bukowski, D. Esau, A. A. Rafat Said, A. Brandt-Talbot and J. Albert, *ChemPlusChem*, 2020, **85**, 373–386.
58. A. Bukowski, K. Schnepf, S. Wesinger, A. Brandt-Talbot and J. Albert, *ACS Sustainable Chem. Eng.*, 2022, **10**, 8474–8483.
59. B. Bertleff, J. Claußnitzer, W. Korth, P. Wasserscheid, A. Jess and J. Albert, *ACS Sustainable Chem. Eng.*, 2017, **5**, 4110–4118.
60. I. V. Babich and J. A. Moulijn, *Fuel*, 2003, **82**, 607–631.
61. B. Bertleff, J. Claußnitzer, W. Korth, P. Wasserscheid, A. Jess and J. Albert, *Energy Fuels*, 2018, **32**, 8383–8388.
62. B. Bertleff, R. Goebel, J. Claußnitzer, W. Korth, M. Skiborowski, P. Wasserscheid, A. Jess and J. Albert, *ChemCatChem*, 2018, **10**, 4602–4609.
63. T. Esser, M. Huber, D. Voß and J. Albert, *Chem. Eng. Res. Des.*, 2022, **185**, 37–50.
64. M. Ahmadian and M. Anbia, *Energy Fuels*, 2021, **35**, 10347–10373.
65. B. Bertleff, M. S. Haider, J. Claußnitzer, W. Korth, P. Wasserscheid, A. Jess and J. Albert, *Energy Fuels*, 2020, **34**, 8099–8109.
66. M. S. Huber, M. Poller, J. Tochtermann, W. Korth, J. Andreas and J. Albert, *Chem. Commun.*, 2023, 4079–4082.
67. M. J. Poller, S. Bönisch, B. Bertleff, J.-C. Raabe, A. Görling and J. Albert, *Chem. Eng. Sci.*, 2022, **264**, 118143.
68. A. F. Holleman, E. und N. Wiberg and G. Fischer, in *Lehrbuch Der Anorganischen Chemie*, Berlin, New York, 2009.
69. B. Weber, in *Koordinationschemie*, Springer, Berlin, Heidelberg, 2014.
70. C. Venturello, R. D'Aloisio, J. C. J. Bart and M. Ricci, *J. Mol. Catal.*, 1985, **32**, 107–110.
71. C. Venturello, E. Alneri and M. Ricci, *J. Org. Chem.*, 1983, **48**, 3831–3833.
72. R. Prabhakar, K. Morokuma, C. L. Hill and D. G. Musaev, *Inorg. Chem.*, 2006, **45**, 5703–5709.
73. A. M. Khenkin, I. Efremenko, J. M. L. Martin and R. Neumann, *J. Am. Chem. Soc.*, 2013, **135**, 19304–19310.
74. J.-M. Brégeault, *Dalton Trans.*, 2003, **3**, 3289–3302.
75. O. Branytska, L. J. W. Shimon and R. Neumann, *Chem. Commun.*, 2007, 3957.
76. V. Kogan, Z. Aizenshtat and R. Neumann, *New J. Chem.*, 2002, **26**, 272–274.
77. A. M. Khenkin, I. Efremenko, L. Weiner, J. M. L. Martin and R. Neumann, *Chem. – Eur. J.*, 2010, **16**, 1356–1364.
78. A. Chemseddine, C. Sanchez, J. Livage, J. P. Launay and M. Fournier, *Inorg. Chem.*, 1984, **23**, 2609–2613.
79. M. D. Tzirakis, I. N. Lykakis and M. Orfanopoulos, *Chem. Soc. Rev.*, 2009, **38**, 2609.
80. S. Esposti, D. Dondi, M. Fagnoni and A. Albini, *Angew. Chem., Int. Ed.*, 2007, **46**, 2531–2534.
81. M. D. Tzirakis and M. Orfanopoulos, *J. Am. Chem. Soc.*, 2009, **131**, 4063–4069.
82. C. Spiteri and J. E. Moses, *Angew. Chem.*, 2010, **122**, 33–36.
83. K. Kamata, Y. Nakagawa, K. Yamaguchi and N. Mizuno, *J. Am. Chem. Soc.*, 2008, **130**, 15304–15310.
84. I. Boldini, G. Guillemot, A. Caselli, A. Proust and E. Gallo, *Adv. Synth. Catal.*, 2010, **352**, 2365–2370.
85. M. Sankar, N. H. Tarte and P. Manikandan, *Appl. Catal., A*, 2004, **276**, 217–222.

6 Applications of Polyoxometalates in Biomedicine

In addition to their catalytic potential, POMs show antiviral and antibacterial activity, as well as antitumor properties and other therapeutic effects. Different tunable characteristics in POM chemistry ensure that POMs can act as optimal therapeutics:[1]

- shape and size of the cluster;
- surface distribution of charge;
- RedOx potential and acidity;
- covalent bonding of organic groups (higher compatibility under physiological conditions); and
- incorporation of POMs in biomacromolecular matrices.

However, POMs show certain limitations when they come into contact with media:[1]

- Under physiological pH values, some POM salts are poorly soluble in water and/or are thermodynamically unstable, resulting in their transformation to other structure types or complete decomposition.
- The reported biological activity of POMs refers to *in vitro* studies and there is a lack of relevant data related to their toxicity in *in vivo* studies.

RSC Foundations No. 3
Polyoxometalate Chemistry
By Jan-Christian Raabe

Published by the Royal Society of Chemistry, www.rsc.org

The following therapeutic effects of POMs are discussed in the literature:[1]

- inhibition of enzymes and proteins;
- inhibition of binding and/or penetration of the virus;
- inhibition of transcription and translation of the deoxyribonucleic acid (DNA); and
- apoptosis (cell death) and/or oxidation of cellular compounds (for antitumor activity).

Note:

- *in vitro* means processes that take place in an artificial environment (*e.g.* a test tube), in general outside living organisms; and
- *in vivo* means processes that take place in the organism.

Important for the understanding of this chapter is the construction of nucleotides, the building blocks for DNA and RNA (ribonucleic acid).[2]

- Nucleotides are built of PO_4^{3-}, a sugar (desoxyribose) and a base (pyrimidine/purine base).[2]
- The bases are adenine and guanine (purine bases), as well as thymine and cytosine (pyrimidine bases).[2]
- Desoxyribose-based nucleotides are the building blocks for DNA and ribose-based nucleotides are the building blocks for RNA.[2]
- Substituting the hydroxyl group of the sugars on C5 with PO_4^{3-} and on C1 with a base yields the nucleotides (deoxyribonucleotides/ribonucleotides), the building blocks of DNA/RNA.[2]
- If nucleotides with different bases are polymerized, a strand is obtained. The polymerization takes place at the PO_4 group (on C5) with a hydroxyl group on C3 of another nucleotide, under elimination of water.[2]
- The single-strand built by ribose-containing nucleotides is called single-stranded RNA (ssRNA). A strand built up by desoxyribose-containing nucleotides is called ssDNA.[2]
- A ssDNA forms a dimer of two strands in which two bases pair *via* hydrogen bonds (base pairing). Adenine pairs with thymine and guanine pairs with cytosine. The dimerized molecule is called DNA or double-stranded DNA. When talking about DNA, usually double-stranded DNA is meant.[2]
- The different single- and double-stranded DNA/RNA polymers differ in their corresponding sequence of nucleotides that contain different bases. The DNA/RNA sequence is also referred to as a base sequence.[2]

Figure 6.1 shows the molecular structures of nucleotides, nucleobases of both sugars, and the structure of ssRNA and DNA (with base pairing).[2]

The genes of every cellular organism consist of DNA, including those of viruses. However, there are also viruses known whose genome consist of RNA, so-called RNA viruses. A virus is a genetic element packaged in a protein envelope that passes from one cell to the next and is unable to reproduce independently.[2]

Biological cells can react to external stimuli (signal–transducer system). This happens if specific molecules interact with the cell membrane embedded proteins, inducing a change in the three-dimensional structure of the proteins. Like a chain reaction, a signal is transmitted inside the cell, causing a reaction of the cell. Due to the interaction of a POM with such membrane proteins, the change in the three-dimensional structure may not be sufficient to trigger the signal, resulting in a missing cell response.[1,2]

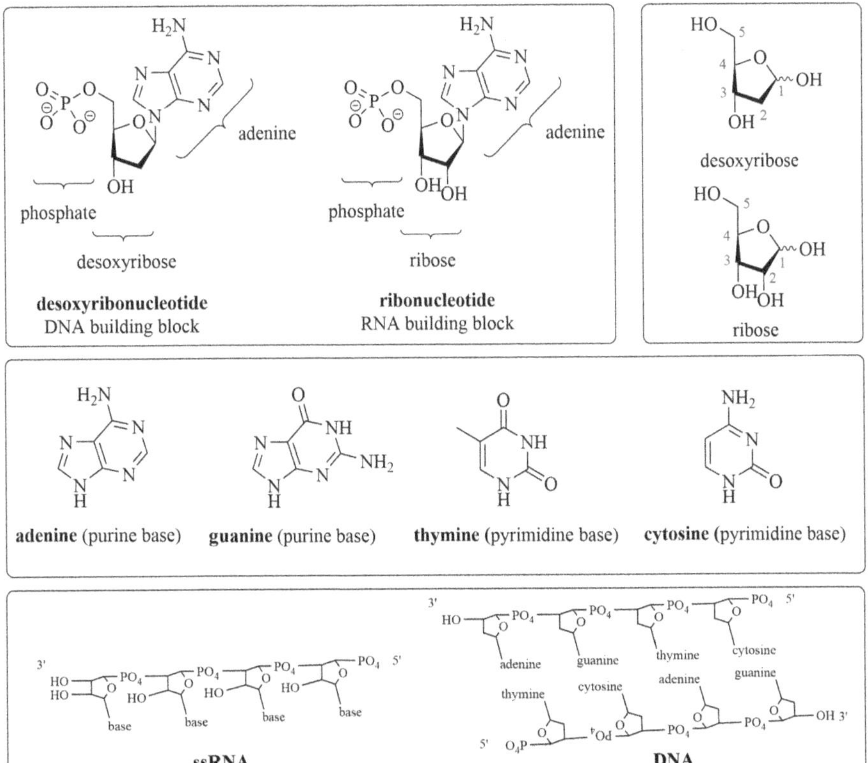

Figure 6.1 Structure of nucleotides, an overview of the different nucleobases and schematic molecular structures of ssRNA and DNA.

However, cell membrane proteins can also act as transporters or transport proteins that transport specific molecules from the outside to the inside of the cell or *vice versa*. In simple words, special activation molecules can bind to those proteins, resulting in a structural change of the protein, making the transporter permeable for the molecule that needs to be transported. POMs can also bind to such transporters *via* ionic interactions and mimic the required activation molecules. A similar effect plays a role in the treatment of diabetes with POMs.[1,2]

6.1 Interactions of Polyoxometalates with Proteins

Finally, this chapter discusses the specific interactions between POMs and proteins. These mechanisms form the basis for the inhibition of enzymes, the initiation of signaling cascades, and the activation of transporters by POMs. There are 20 proteinogenic amino acids relevant for the formation of proteins, which are assembled in a defined sequence to form a biomacromolecule with a defined three-dimensional structure. The interactions between proteins and POMs were investigated using the $[V_{10}O_{28}]^{6-}$ structure as an example.[3]

The most important non-covalent interactions between POMs and proteins are (Figure 6.2):

- Hydrogen bonds. These interactions are found between the side chains of the amino acid serine and the terminal oxygen atoms $V{=}O_t$. External mediator molecules (*e.g.* a water molecule) can also be used to form hydrogen bonds to the POM.[3]
- Electrostatic/ionic interactions. Purely electrostatic interactions are found between the POM and the cationic side chains of the amino acids arginine or lysine, compensating the high anionic charge of a POM cluster.[3]
- Coordinative interactions. Coordinative interactions are found between the anionic side chains of aspartic acid and glutamic acid, which coordinate to a cation. This cation can coordinate to the terminal oxo ligands of a POM. Electrostatic interactions are also found, *e.g.* if a more highly charged cation coordinates to the carboxylate functionality, which interacts electrostatically with the POM.[3]

Due to the specific interactions between POMs and proteins, POMs are interesting candidates to help with the crystallization of proteins for the following reasons:[4]

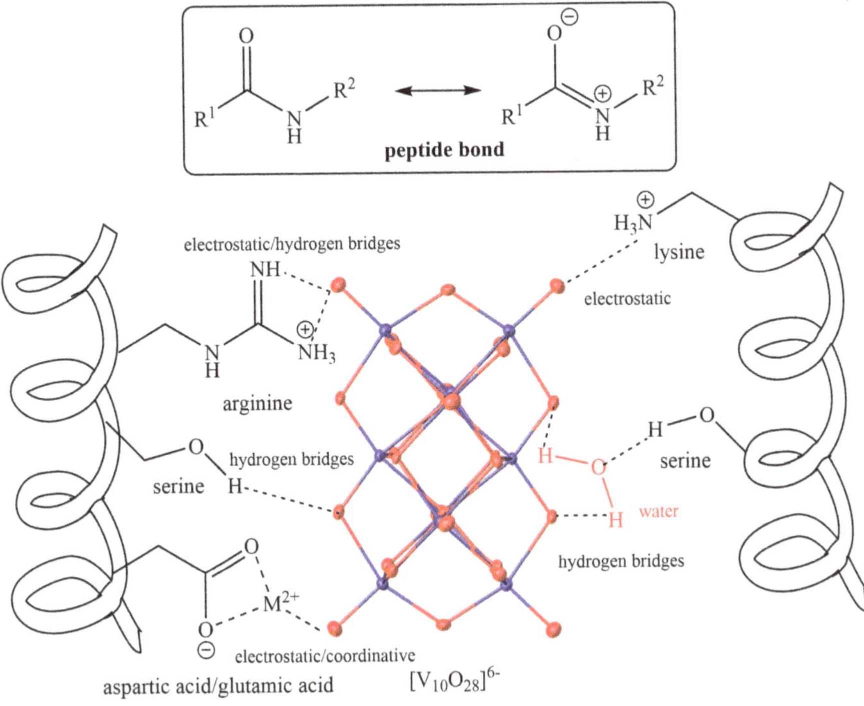

Figure 6.2 Interaction of POMs (here $[V_{10}O_{28}]^{6-}$) with amino acid side chains in a protein.

- Solving the phase problem is simplified because heavy atoms are present.[4]
- Due to their specific interactions with the protein, POMs can help the protein to crystallize faster.[4]
- Due to their specific interactions with the protein, POMs can help the protein to crystallize in a defined conformation.[4]
- Due to their specific interactions with the protein, POMs can stabilize the crystal and help with crystal packing. The POMs can cross-link the proteins, which promotes stability and fast crystal packing.[4]

6.2 Applications of Polyoxometalates as Antiviral Agents

6.2.1 Viral Diseases

The human immunodeficiency virus (HIV) is one of the biggest health threats in the world. In general, there are two types of HIV: HIV-1 and HIV-2. HIV-1 is the most common type of infection and when talking about "HIV" usually HIV-1 is being referred to. HIV-2 occurs in a much smaller number of people. Infections are found mostly in West Africa and make up ~0.01% of all HIV infections. For both types, there are multiple groups and subgroups known.[5–7]

T lymphocytes, or T cells, are a group of white blood cells that serve the immune defense. $CD4^+$ cells form a subgroup that is attacked by HIV, which leads to the outbreak of acquired immune deficiency syndrome (AIDS).[5,8] The glycoprotein gb120 is exposed on the surface of the HIV envelope. The 120 in its name means that the molar mass is 120 kDa. This protein plays a significant role in the entry of the virus into cells as it can attach to specific cell surface receptors. It can interact with the receptors of the $CD4^+$ cells, starting a cascade of conformational changes in the gb120 protein. As a result, this leads to fusion of the HIV viral membrane with the host cell membrane of $CD4^+$. The fusion of cells is called syncytium. The binding to $CD4^+$ cells is mainly comprised of electrostatic interactions.[9–12] This means that POMs as polyanionic clusters can interact with those cells and can block the binding of HIV to $CD4^+$ cells.[13]

Hepatitis is a viral infection of the liver. The disease can be divided into different forms of hepatitis, A to E.[14] More than 70% of all infections result in a persistent infection (liver cirrhosis, hepatocellular carcinoma). About 2% of the human world population is infected with hepatitis.[15]

The half-maximal inhibitory concentration (IC_{50}) is the most effective and informative measure of the efficacy of drugs. This value indicates how much of a drug is needed for a half inhibition of a biological process.[16] Another measure is the half maximal effective concentration (EC_{50}), which is a drug concentration that induces a biological response halfway between the baseline and the maximum after a defined time of exposure. It is defined as the concentration that is needed to obtain 50% of the effect.[17]

6.2.2 Antiviral Properties of Polyoxometalates

HIV-1(III_b), HIV-1(ROD) and HIV-2(ROD) represent different subgroups within the virus types HIV-1 and HIV-2. The effect of different monomeric, dimeric and trimeric Keggin- and Wells–Dawson-type POMs on the replication of HIV-1(III_b) and HIV-1(ROD) was investigated in detail. HIV-2(ROD) becomes less sensitive with an increasing number of Keggin- and Wells–Dawson-type structures per molecule, meaning that HIV-2(ROD) is sensitive towards monomeric Keggin-type POMs ($IC_{50} < 4\ \mu g\ mL^{-1}$ monomeric structures and $IC_{50} > 71.8\ \mu g\ mL^{-1}$ trimeric structures). The antiviral activity against

HIV-1(III_b) of mono-, di- and trimeric Keggin-type POMs is comparable (IC_{50} 0.4–0.5 $\mu g\,mL^{-1}$).[1,13] The dimeric Wells–Dawson-type structure $Na_{16}[Mn_4(H_2O)_2(P_2W_{15}O_{56})_2]$ blocks the binding of gb120 to SUP-T1 cells (T cells) and therefore has a high antiviral activity (IC_{50} 1.2 $\mu g\,mL^{-1}$ for HIV-1(III_b) and 13.9 $\mu g\,mL^{-1}$ for HIV-2(ROD)). There are some differences in the inhibition effects against HIV-1 and HIV-2, suggesting structural and conformational differences in the target protein molecules with which POMs can interact.[1,13]

Promising anti-RNA virus activity was found for Keggin-type polyoxotungstates substituted with Ti and V, in both *in vivo* and *in vitro* cases. Potent and broad antiviral activity was found against viruses causing acute respiratory infections and against HIV-1. Keggin-type species, *e.g.* $(^iPrNH_3)_6[PTi_2W_{10}O_{38}(O_2)_2]$ (EC_{50} 5.6 μM) and the sandwich Keggin-type species $K_{10}Na[(V^{IV}O)_2(V^{V}O)(SbW_9O_{33})_2]$ (EC_{50} 1.75 μM), and POM $K_{11}H[(V^{V}O)_3(SbW_9O_{33})_2]$ (EC_{50} 4.0 μM), show potent anti-influenza virus (FluV A) activity *in vitro*. POM $(^iPrNH_3)_6[PTi_2W_{10}O_{38}(O_2)_2]$ is a Ti(IV) substituted Keggin-type species in which two oxo ligands have been replaced by peroxido ligands. The cations are iso-propylammonium ions. POM $K_{10}Na[(V^{IV}O)_2(V^{V}O)(SbW_9O_{33})_2]$ is a sandwich-type POM in which the lacunary-type species (SbW_9O_{33}) coordinates to two $V^{IV}O^{2+}$ (oxidation state +4) and one $V^{V}O^{3+}$ (oxidation state +5) cations. In POM $K_{11}H[(V^{V}O)_3(SbW_9O_{33})_2]$, all vanadium atoms are present in oxidation state +5.[1,18,19] Against FluV A virus, the EC_{50} value for POM $(^iPrNH_3)_6[PTi_2W_{10}O_{38}(O_2)_2]$ is 2.4 μM.[1,18] All POMs also show inhibitory effects against HIV replication, with EC_{50} values under 2.0 μM. The V(V) and V(V)/V(IV) mixed substituted POMs show the highest inhibitory effect with a low toxicity. POM $K_{10}Na[(V^{IV}O)_2(V^{V}O)(SbW_9O_{33})_2]$ can bind to gp120 and interfere with the interaction between gp120 and its receptor, avoiding syncytium formation.[1,19] Information about the relationship between POM structure type *versus* antiviral activity is rare. It is assumed that water-soluble POMs of the Wells–Dawson-type structure show the highest inhibitory effect on HIV-1, whereas Keggin-type structures show a significantly lower inhibitory effect. The introduction of organic ligands boosts the antiviral activity, possibly due to an increased lipophilicity.[1,20]

Monomeric Keggin-type POMs, especially $Cs_2K_4Na[SiW_9Nb_3O_{40}]$, show an inhibition effect against hepatitis C virus (HCV) infections with a low cytotoxicity (EC_{50} 0.8 μM). This POM can break down the HCV envelope and disrupt the integrity of the viral particles.[1,15]

It can be concluded that clinically approved drugs are still the best option for fighting against viruses. However, POM-based approaches show very promising results against viruses and can perhaps be used as therapeutics in the future. An ongoing problem is the toxicity of the metal-based clusters, which must be overcome before POMs can be considered as real and cheap inorganic antiviral agents in medicine.[1]

6.3 Applications of Polyoxometalates as Antibacterial Agents

6.3.1 Multi-resistant Germs

Pathogenic bacteria cause many infections and are one of the biggest health problems facing humanity in the 21st century. The administration of antibiotics can positively influence the course of an infection and bring about a rapid recovery. However, the routine use of antibiotics leads to the development of multi-resistant germs. This is due to so-called resistance mutations in the genome of bacteria. The constant development and marketing of new antibiotics is leading to the development of further resistance, particularly due to the improper use of newly developed antibiotics, causing the therapeutic effect of antibiotics to suffer.[21–23]

A well-known resistant bacterium is methicillin-resistant *Staphylococcus aureus* (MRSA), which is the most clinically relevant and is found in the human nasal mucosa of 20% to 40% of the population. This type of bacteria was first reported in England in 1961, after the antibiotic methicillin was used in clinical medical practice. It was responsible for many outbreaks in hospitals in many parts of the world, especially in the 1980s (Australia), 1990s (United States of America) and in the 2000s (livestock exposure).[24] Vancomycin-resistant *Staphylococcus aureus* (VRSA) is another resistant type of bacteria. The antibiotic vancomycin has been used in the fight against MRSA. In 1996, the first vancomycin-resistant MRSA was isolated in Japan from a patient who contracted a wound infection that was treated with vancomycin. Later it was shown that VRSA is a global issue.[25]

Moraxella catarrhalis is found in the upper respiratory tract and has emerged as a pathogen over the last 20 to 30 years. It is an important cause of upper respiratory tract infections and of lower respiratory tract infections, especially in adults with chronic obstructive pulmonary disease (COPD). In immunocompromised hosts, it is responsible for a variety of severe infections like pneumonia, endocarditis (inflammation of the inner lining of the heart), septicemia (blood poisoning), and meningitis (inflammation of the meninges).[26]

6.3.2 Antibacterial Properties of Polyoxometalates

Different POMs were reported to show activity against Gram-positive and -negative bacteria, showing an antibiotic effect.[1]

The Keggin- and Wells–Dawson-type clusters $K_7[PTi_2W_{10}O_{40}]$, $K_4[SiMo_{12}O_{40}]$ and $K_6[P_2W_{18}O_{62}]$ are known to improve antibacterial effectiveness against MRSA and VRSA in synergy with β-lactam antibiotics. Production of penicillin-binding protein 2′ (PBP2′) is known to be associated with β-lactam-resistance for MRSA and VRSA. With the above-mentioned POMs, PBP2′ production can be inhibited, increasing the sensitivity of MRSA and VRSA to β-lactam antibiotics. POMs interact with the cell membrane proteins, which are related to the signal–transducer system, to produce PBP2′. This interaction suppresses the formation of the PBP2′ protein, which makes the bacterium sensitive to the administered antibiotic.[1,27–29]

The Preyssler-type anion $[NaP_5W_{30}O_{110}]^{14-}$ is known for its activity against *Moraxella catarrhalis*. The Wells–Dawson-type anion $[P_2W_{18}O_{62}]^{6-}$ and the Keggin-type species $[H_2CoTiW_{11}O_{40}]^{6-}$ show comparable activity. Here, the antibacterial activity correlates with the composition, shape and size of the cluster. For POMs with anionic charges of more than 12 and framework elements ≤ 22, the activity correlates with the charge of the cluster. It is likely that POM clusters can be absorbed onto the cellular membrane of the bacteria *via* electrostatic and/or hydrophobic interactions. The strength of this interaction depends on the charge of the lipid layer.[30]

To reduce the toxicity of POMs, modification with organic ligands or encapsulation in non-toxic macromolecular matrices is possible. An example of the encapsulation of POMs is the modification of POMs with chitosan, a biopolymer. Chitosan nanoparticles are easy to produce and the release of the drug under physiological conditions is very simple, as chitosan can be degraded by enzymes such as chitosanase or lysozyme. However, chitosan is only soluble in acidic aqueous media. This can be circumvented by modifying the chitosan with water-soluble carboxymethyl chitosan (CMC) by carboxymethylation of the alcohol and amino groups. The resulting nanocomposites have a size between 50 and 160 nm and are therefore ideally sized to be taken up by the cells (see Figure 6.3).[31]

Another approach is to modify the POM with organic ligands. This allows specific control of the shape, size, lipophilicity, solubility, stability, RedOx and acid/base properties of the clusters. One example is the sandwich-type POM $(NH_4)_{12}[(PhSb)_4(GeW_9O_{34})_2]$, in which the

Chitosan

R = H **Chitosan**

R = CH_2COOH, CH_2COONa **CMC**

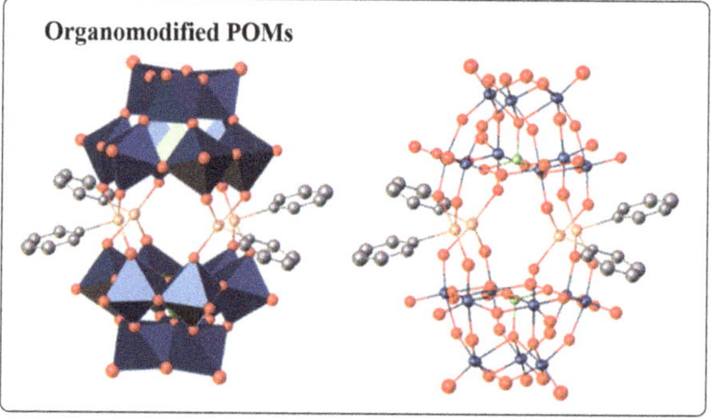

$$2\,Na_{10}[GeW_9O_{34}] + 4\,PhSbCl_2 + 12\,NH_4Cl \xrightarrow[4\text{-}5\,^{\circ}C,\ 1\ day]{H_2O,\ 30\ min,\ 20\,^{\circ}} (NH_4)_{12}[(PhSb)_4(GeW_9O_{34})_2] + 20\,NaCl$$

Figure 6.3 Chemical structure of the chitosan polymer (top) and $(NH_4)_{12}[(PhSb)_4(GeW_9O_{34})_2]$ (middle) and a synthetic reaction equation (bottom). The three-dimensional structure of the anion is shown in its polyhedron (left) and atomistic representation (right). Color code: green – Ge, blue – W, red – O, orange – Sb and gray – C. Hydrogen atoms are not shown. The data were used from the Cambridge Crystallographic Data Centre and Fachinformationszentrum Karlsruhe Access Structures service database (deposition number: 891299).

two Keggin lacunary-type structures $[GeW_9O_{34}]^{10-}$ (germanotungstate) coordinate to four $[SbPh]^{2+}$ units. The POM is prepared from $PhSbCl_2$ and the lacunary-type anions according to the scheme shown in Figure 6.3. The compound shows antibacterial properties caused by selective bacterial cell toxicity of the metals after the POM has been taken up into the bacterial cells.[32,33]

6.4 Applications of Polyoxometalates as Antitumor Agents

Cancer is one of the leading causes of death in the world. The most important challenge in the fight against cancer is to stop the growth of cancer cells. Chemotherapy is the most widely used method, but the drugs show a lack of selectivity, causing many unwanted side effects. Therefore it is necessary to search for new agents that can limit the growth of cancer cells.[34]

Xenografting is the transfer of functional, living cells between different species. This technique was used to transfer human colon, breast and lung cancer cells to athymic nude mice in order to study the antitumor activity of the paramolybdate anion *in vivo*. The antitumor activity of the isopropylammonium salt $(^{i}PrNH_3)_6[Mo_7O_{24}]$ is comparable to reference antitumor drugs.[1,35] In other studies on human pancreatic cancer cells, the antitumor effect was attributed to the initiation of an apoptotic pathway (initiation of programmed cell death).[1,36,37]

Ecto-nucleotidases are a family of nucleotide metabolizing enzymes, which are expressed in the cell membrane. Nucleotides can be metabolized to nucleosides by these enzymes by splitting off the PO_4^{3-} group. The remaining molecule, which consists only of the sugar and base, is called a nucleoside, as shown in Figure 6.4.[2,38] This type of enzyme is involved in adenosine formation, an important molecule for cell growth. Cell growth is a major characteristic of all malignant tumors and depends on the availability and synthesis of nucleosides.[39] It was shown that the Co(II)-containing sandwich-type cluster $[Co_4(H_2O)_2(PW_9O_{34})_2]^{10-}$ is a human ecto-nucleotidase inhibitor in the nano- and micromolecular range. Specifically, it is an ecto-nucleoside triphosphate diphosphohydrolases (NTPDase) inhibitor. The NTPDase catalyzes the cleavage of a PO_4^{3-} group of a nucleotide triphosphate, yielding a nucleotide diphosphate, as shown in Figure 6.4.[1,40,41]

POM acids, like $H_4[SiW_{12}O_{40}]$ and $H_4[SiMo_{12}O_{40}]$, are also known to act as ecto-nucleotidase inhibitors and have been tested on rat synaptic plasma membranes. The pH values of these enzyme assays are about 7.4, meaning that these POMs are transformed into their mono lacunary-type species $[SiW_{11}O_{39}]^{8-}$ and $[SiMo_{11}O_{39}]^{8-}$ under physiological conditions.[1,42]

For the decavanadate anion, a cation dependency was found. Salts of the type $(H_2tmen)_3[V_{10}O_{28}]$ (H_2tmen means double protonated *N,N,N′,N′*-tetramethylethylenediammonium) and $(H_2en)_3[V_{10}O_{28}]$ (H_2en means double protonated ethylendiamine) were tested against murine leukemia and human lung carcinoma cells. The more potent inhibitory effect of salt $(H_2tmen)_3[V_{10}O_{28}]$ is attributed to the higher lipophilic character of the cation, which can pass the cell membrane together with the POM anion.[1,43]

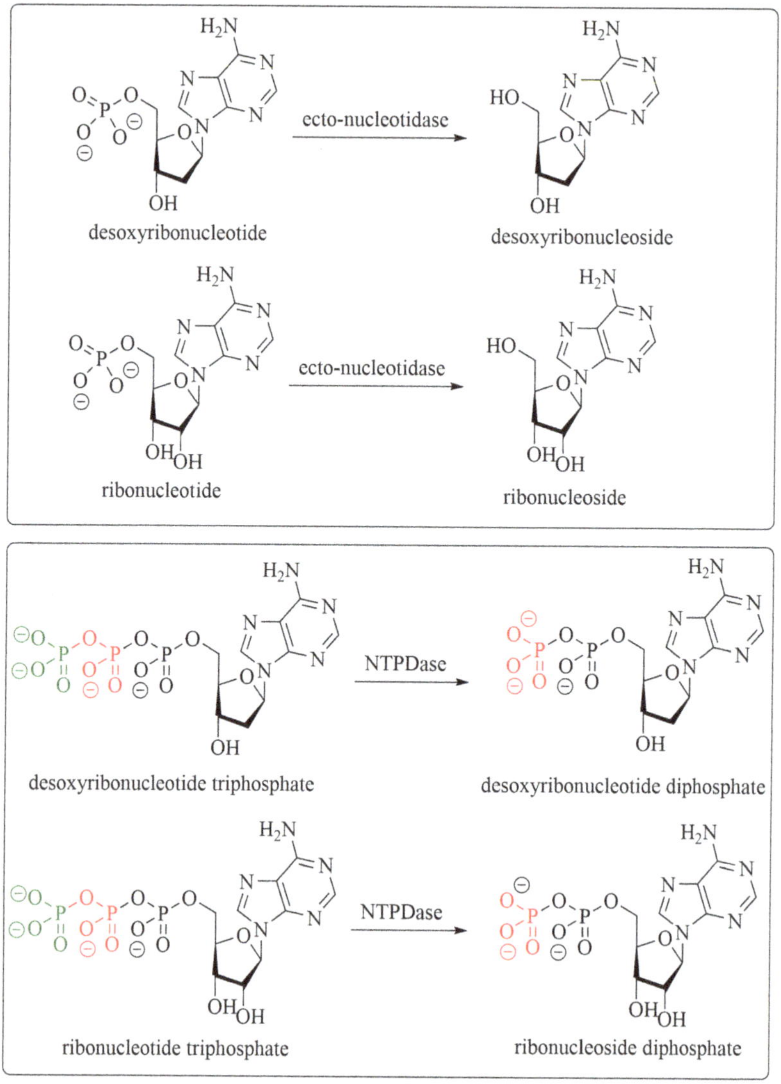

desoxyribonucleotide — ecto-nucleotidase → desoxyribonucleoside

ribonucleotide — ecto-nucleotidase → ribonucleoside

desoxyribonucleotide triphosphate — NTPDase → desoxyribonucleotide diphosphate

ribonucleotide triphosphate — NTPDase → ribonucleoside diphosphate

Figure 6.4 The enzyme ecto-nucleotidase catalyzes the cleavage of the PO_4^{3-} group from the nucleotide (deoxyribonucleotide/ribonucleotide). A nucleoside (deoxyribonucleoside/ribonucleoside) remains.

Cell growth suppression induced by POMs is higher than that observed with commonly used cytostatic drugs. However, their use in clinical practice is extremely challenging due to their cytotoxicity and unspecific interactions with biomacromolecules.[34]

6.5 Applications of Polyoxometalates as Antidiabetic Agents and Normoglycemic Properties

Diabetes, also known as diabetes mellitus, is a metabolic disorder that leads to hyperglycemia (increased blood sugar concentrations) in the blood due to a lack of insulin. Insulin is a hormone that regulates blood sugar levels by stimulating the body to absorb sugar from the blood. In 2011, there were an estimated 366 million sufferers, a number that is expected to rise to 552 million by 2030. A normal blood sugar concentration is called normoglycemia.[44] Insulin is known to bind to an insulin receptor, a cell membrane protein that is embedded in the cell membrane. This interaction starts a protein activation cascade, leading the glucose transporter membrane protein to take up the glucose into the cell.[45]

It is assumed that POMs can also bind to the insulin receptors, whereby the POM mimics the insulin molecule and initiates the same protein activation cascade, causing glucose to be taken up into the cell by the transport protein. The Wells–Dawson-type anions $[P_2W_{18}O_{62}]^{6-}$, $[P_2VW_{17}O_{62}]^{7-}$ and $[P_2V_3W_{15}O_{62}]^{9-}$, as well as the decavanadate anion $[V_{10}O_{28}]^{6-}$, showed an *in vivo* hypoglycemic effect (blood sugar concentration lowering effect), which was attributed to the insulin-mimetic effect related to the structural features of a Wells–Dawson-type structure.[1,46] For $[V_{10}O_{28}]^{6-}$ in concentrations of about 1 mM, an increased glucose uptake was observed, an additive effect that increases glucose uptake by as much as 50% compared to glucose uptake induced by insulin alone.[1,47]

6.6 Applications of Polyoxometalates as Potential Drugs for the Treatment of Alzheimer's Disease

Alzheimer's disease (AD) is one of the most common causes of dementia in older people and is associated with memory loss, emotional disturbances and personality changes. Current models assume that the loss of memory is associated with a decrease in the concentration of the neurotransmitter acetylcholine in the hippocampus and cortex. Drugs that increase the concentration of acetylcholine, such as cholinesterase inhibitors, are therefore used for treatment.[48,49] Another approach is to develop inhibitors for the enzymes acetylcholinesterase

Cholinesterase/AChE/BChE
- AA

Scheme 6.1 Cholinesterase/AChE/BChE-catalyzed ester hydrolysis of acetylcholine.

(AChE) and butyrylcholinesterase (BChE). AChE and BChE are enzymes that can hydrolyze choline esters, such as acetylcholine, into AA and choline (Scheme 6.1).[50,51] Together with AChE, BChE is an essential component in the metabolism of neurotransmitters.[51] The Anderson–Evans type POM $[TeW_6O_{24}]^{6-}$ and the anion $[H_2W_{12}O_{42}]^{10-}$ proved to be selective inhibitors of these enzymes with IC_{50} values in the range of 0.31 ± 0.01 and 0.29 ± 0.01 μM.[48]

6.7 Concluding Remarks

The applications of POMs in biomedicine that have been discussed in this chapter show their valuable potential in the treatment of diseases that threaten human health in the 21st century. However, it must be noted that transition metals show a high toxicity, so POMs must be developed where the toxicity is kept to a minimum. In addition to the identification of medically effective POMs, future studies need to focus on the underlying mechanisms in order to tailor POMs specifically for medical uses.

Abbreviations

POM	Polyoxometalate
DNA	Deoxyribonucleic acid
RNA	Ribonucleic acid
ss	Single-stranded
HIV	Human immunodeficiency virus
AIDS	Acquired immune deficiency syndrome
IC_{50}	Half-maximal inhibitory concentration
EC_{50}	Half maximal effective concentration
FluV	Influenza virus
HCV	Hepatitis C virus
MRSA	Methicillin-resistant *Staphylococcus aureus*
VRSA	Vancomycin-resistant *Staphylococcus aureus*
COPD	Chronic obstructive pulmonary disease
PBP2′	Penicillin-binding protein 2′
CMC	Carboxymethyl chitosan

RedOx	Reduction/oxidation
NTPDase	Ecto-nucleoside triphosphate diphosphohydrolases
pH	Negative decadic logarithm of the hydrogen ion concentration
H_2tmen	Double protonated *N*,*N*,*N*′,*N*′-tetramethylethylene-diammonium
H_2en	Double protonated ethylendiamine
AD	Alzheimer's disease
AChE	Acetylcholinesterase
BChE	Butyrylcholinesterase
AA	Acetic acid

Acknowledgements

I would like to thank the publisher, the Royal Society of Chemistry, for the opportunity to write this book!

Recommended Reading

Please have a look into the following references:

Overview of polyoxometalates in biomedicine:

1. M. B. Čolović, M. Lacković, J. Lalatović, A. S. Mougharbel, U. Kortz and D. Z. Krstić, Med, *Curr. Med. Chem.*, 2019, **27**, 362–379.

Interactions of polyoxometalates with proteins:

1. M. Aureliano, N. I. Gumerova, G. Sciortino, E. Garribba, C. C. McLauchlan, A. Rompel and D. C. Crans, *Coord. Chem. Rev.*, 2022, **454**, 214344.

Basics of biochemistry:

1. J. M. Berg, J. L. Tymoczko, G. J. Gatto and L. Stryer, *Biochemistry*, New York, 9th edn, 2019.

References

1. M. B. Čolović, M. Lacković, J. Lalatović, A. S. Mougharbel, U. Kortz and D. Z. Krstić, *Med, Curr. Med. Chem.*, 2019, **27**, 362–379.
2. J. M. Berg, J. L. Tymoczko, G. J. Gatto and L. Stryer, in *Stryer Biochemie*, Springer, Berlin, Heidelberg, 2018.
3. M. Aureliano, N. I. Gumerova, G. Sciortino, E. Garribba, C. C. McLauchlan, A. Rompel and D. C. Crans, *Coord. Chem. Rev.*, 2022, **454**, 214344.
4. A. Bijelic and A. Rompel, *Coord. Chem. Rev.*, 2015, **299**, 22–38.
5. R. Seitz, *Transfus. Med. Hemother.*, 2016, **43**, 203–222.

6. J. Coffin, A. Haase, J. A. Levy, L. Montagnier, S. Oroszlan, N. Teich, H. Temin, K. Toyoshima, H. Varmus, P. Vogt and R. Weiss, *Science*, 1986, **80**, 697.
7. J. D. Reeves and R. W. Doms, *J. Gen. Virol.*, 2002, **83**, 1253–1265.
8. R. V. Luckheeram, R. Zhou, A. D. Verma and B. Xia, *Clin. Dev. Immunol.*, 2012, 1–12.
9. J. Sodroski, R. Patarca, D. Perkins, D. Briggs, T.-H. Lee, M. Essex, J. Coligan, F. Wong-Staal, R. C. Gallo and W. A. Haseltine, *Science*, 1984, **80**, 421–424.
10. B. M. Curtis, S. Scharnowske and A. J. Watson, *Proc. Natl. Acad. Sci. U. S. A.*, 1992, **89**, 8356–8360.
11. L. de Witte, M. Bobardt, U. Chatterji, G. Degeest, G. David, T. B. H. Geijtenbeek and P. Gallay, *Proc. Natl. Acad. Sci. U. S. A.*, 2007, **104**, 19464–19469.
12. A. G. Dalgleish, P. C. L. Beverley, P. R. Clapham, D. H. Crawford, M. F. Greaves and R. A. Weiss, *Nature*, 1984, **312**, 763–767.
13. M. Witvrouw, H. Weigold, C. Pannecouque, D. Schols, E. De Clercq and G. Holan, *J. Med. Chem.*, 2000, **43**, 778–783.
14. J. A. Cuthbert, *Clin. Microbiol. Rev.*, 2001, **14**, 38–58.
15. Y. Qi, Y. Xiang, J. Wang, Y. Qi, J. Li, J. Niu and J. Zhong, *Antiviral Res.*, 2013, **100**, 392–398.
16. S. Aykul and E. Martinez-Hackert, *Anal. Biochem.*, 2016, **508**, 97–103.
17. Z. Chen, R. Bertin and G. Froldi, *Food Chem.*, 2013, **138**, 414–420.
18. S. Shigeta, S. Mori, T. Yamase, N. Yamamoto and N. Yamamoto, *Biomed. Pharmacother.*, 2006, **60**, 211–219.
19. S. Shigeta, S. Mori, E. Kodama, J. Kodama, K. Takahashi and T. Yamase, *Antiviral Res.*, 2003, **58**, 265–271.
20. A. Flütsch, T. Schroeder, M. G. Grütter and G. R. Patzke, *Bioorg. Med. Chem. Lett.*, 2011, **21**, 1162–1166.
21. G. Tangherlini, T. Torregrossa, O. Agoglitta, J. Köhler, J. Melesina, W. Sippl and R. Holl, *Bioorg. Med. Chem.*, 2016, **24**, 1032–1044.
22. H. Kurasaki, K. Tsuda, M. Shinoyama, N. Takaya, Y. Yamaguchi, R. Kishii, K. Iwase, N. Ando, M. Nomura and Y. Kohno, *ACS Med. Chem. Lett.*, 2016, **7**, 623–628.
23. P. Zhou and A. Barb, *Curr. Pharm. Biotechnol.*, 2008, **9**, 9–15.
24. A. S. Lee, H. de Lencastre, J. Garau, J. Kluytmans, S. Malhotra-Kumar, A. Peschel and S. Harbarth, *Nat. Rev. Dis. Primers*, 2018, **4**, 18033.
25. K. Hiramatsu, *Lancet Infect. Dis.*, 2001, **1**, 147–155.
26. C. M. Verduin, C. Hol, A. Fleer, H. van Dijk and A. van Belkum, *Clin. Microbiol. Rev.*, 2002, **15**, 125–144.
27. T. Yamase, *J. Mater. Chem.*, 2005, **15**, 4773.
28. M. Inoue, T. Suzuki, Y. Fujita, M. Oda, N. Matsumoto, J. Iijima and T. Yamase, *Biomed. Pharmacother.*, 2006, **60**, 220–226.
29. J. Fishovitz, J. A. Hermoso, M. Chang and S. Mobashery, *IUBMB Life*, 2014, **66**, 572–577.
30. N. I. Gumerova, E. Al-Sayed, L. Krivosudský, H. Ĉipĉić-Paljetak, D. Verbanac and A. Rompel, *Front. Chem.*, 2018, **6**, 1–9.
31. G. Geisberger, S. Paulus, M. Carraro, M. Bonchio and G. R. Patzke, *Chem. – Eur. J.*, 2011, **17**, 4619–4625.
32. M. Barsukova-Stuckart, L. F. Piedra-Garza, B. Gautam, G. Alfaro-Espinoza, N. V. Izarova, A. Banerjee, B. S. Bassil, M. S. Ullrich, H. J. Breunig, C. Silvestru and U. Kortz, *Inorg. Chem.*, 2012, **51**, 12015–12022.
33. J. Gu, L. Zhang, X. Yuan, Y.-G. Chen, X. Gao and D. Li, *Bioinorg. Chem. Appl.*, 2018, **2018**, 1–6.
34. M. J. W. Budych, K. Staszak, A. Bajek, F. Pniewski, R. Jastrząb, M. Staszak, B. Tylkowski and K. Wieszczycka, *Coord. Chem. Rev.*, 2023, **493**, 215306.
35. H. Fujita, T. Fujita, T. Sakurai, T. Yamase and Y. Seto, *Tohoku J. Exp. Med.*, 1992, **168**, 421–426.

36. A. Ogata, H. Yanagie, E. Ishikawa, Y. Morishita, S. Mitsui, A. Yamashita, K. Hasumi, S. Takamoto, T. Yamase and M. Eriguchi, *Br. J. Cancer*, 2008, **98**, 399–409.
37. A. Ogata, S. Mitsui, H. Yanagie, H. Kasano, T. Hisa, T. Yamase and M. Eriguchi, *Biomed. Pharmacother.*, 2005, **59**, 240–244.
38. G. Guido, *Front. Biosci.*, 2008, **13**, 2588.
39. X. Zhou, X. Zhi, P. Zhou, S. Chen, F. Zhao, Z. Shao, Z. Ou and L. Yin, *Oncol. Rep.*, 2007, 1341–1346.
40. S.-Y. Lee, A. Fiene, W. Li, T. Hanck, K. A. Brylev, V. E. Fedorov, J. Lecka, A. Haider, H.-J. Pietzsch, H. Zimmermann, J. Sévigny, U. Kortz, H. Stephan and C. E. Müller, *Biochem. Pharmacol.*, 2015, **93**, 171–181.
41. C. E. Müller, J. Iqbal, Y. Baqi, H. Zimmermann, A. Röllich and H. Stephan, *Bioorg. Med. Chem. Lett.*, 2006, **16**, 5943–5947.
42. M. B. Čolović, D. V. Bajuk-Bogdanović, N. S. Avramović, I. D. Holclajtner-Antunović, N. S. Bošnjaković-Pavlović, V. M. Vasić and D. Z. Krstić, *Bioorg. Med. Chem.*, 2011, **19**, 7063–7069.
43. Y.-T. Li, C.-Y. Zhu, Z.-Y. Wu, M. Jiang and C.-W. Yan, *Transition Met. Chem.*, 2010, **35**, 597–603.
44. U. Alam, O. Asghar, S. Azmi and R. A. Malik, *Handb. Clin. Neurol.*, 2014, 211–222.
45. A. Krook, W.-H. Wallberg-Henriksson and J. R. Zeirath, *Med. Sci. Sports Exercise*, 2004, **36**, 1212–1217.
46. K. Nomiya, H. Torii, T. Hasegawa, Y. Nemoto, K. Nomura, K. Hashino, M. Uchida, Y. Kato, K. Shimizu and M. Oda, *J. Inorg. Biochem.*, 2001, **86**, 657–667.
47. M. J. Pereira, E. Carvalho, J. W. Eriksson, D. C. Crans and M. Aureliano, *J. Inorg. Biochem.*, 2009, **103**, 1687–1692.
48. J. Iqbal, M. Barsukova-Stuckart, M. Ibrahim, S. U. Ali, A. A. Khan and U. Kortz, *Med. Chem. Res.*, 2013, **22**, 1224–1228.
49. G. Pepeu and M. Giovannini, *Alzheimer's Res.*, 2009, **6**, 86–96.
50. H. Soreq and S. Seidman, *Nat. Rev. Neurosci.*, 2001, **2**, 294–302.
51. S. Darvesh, D. A. Hopkins and C. Geula, *Nat. Rev. Neurosci.*, 2003, **4**, 131–138.

7 Conclusion

POMs represent a fascinating class of oxometal complexes in their d^0 configuration. The easiest oxo complex is the hexaoxometal complex MO_6. It is possible to use POMs in a range of different applications because of their structural diversity. POMs are small inorganic nanoclusters that are formed from transition elements in their highest possible oxidation states. The best-known elements are Mo(VI), W(VI), V(V), Nb(V) and Ta(V), which act as so-called framework elements in their highest oxidation states. Each framework element is linked to the next *via* a so-called oxo ligand bridge and has at least one terminal oxo ligand. Each metal is coordinated octahedrally by six other oxo ligands, with the octahedra linked *via* common edges and corners to form different structure types. Some POMs also contain one or more so-called heteroelements, which are usually main-group elements. Such structures are then classified as HPA structures, while structure types without heteroelements are referred to as IPA structures. Formally, POMs are formed under acidic conditions from molecular monometalates MO_4^{2-} (or from metal oxides), which are converted into POMs by oligomerization under elimination of water. Alternatively, POMs can also be formed by the degradation of metal oxides under basic conditions (*e.g.* $[Nb_6O_{19}]^{8-}$ or $[Ta_6O_{19}]^{8-}$). This is possible because there are no monometalates MO_4^{2-} of Nb(V) and Ta(V) and the basic degradation stops at the level of hexametalates. From a coordination chemical point of view the oxo ligand is a π-donor ligand.

The formation of oxo clusters is driven by the so-called oxophilicity of the respective metals to the element oxygen. Oxo compounds are known for each element in the periodic table, as oxygen forms stable

RSC Foundations No. 3
Polyoxometalate Chemistry
By Jan-Christian Raabe

Published by the Royal Society of Chemistry, www.rsc.org

compounds with every element, *e.g.* water, organic molecules, *etc.* In most cases, oxygen is the element with the negative polarization, which determines the chemistry of these molecules. There is also an affinity between metals and other non-metals, but this is not as pronounced as with oxygen. This also explains the stability of POMs in aqueous media, as the oxygen atom in the water molecule can only hydrolyze the saturated oxygen bridges in the cluster to a limited extent (see dissociation equilibria of POMs in aqueous media). On the other hand, metal complexes with non-metal elements other than oxygen tend to hydrolyze in aqueous media due to the strong oxophilicity of the oxygen atom in water molecules. The chemistry of POMs would probably not be possible if the affinity of metals for oxygen was not pronounced.

POMs are synthesized by two approaches. In the self-assembly approach, the precursors form POM clusters driven by thermodynamic driving forces. If a POM cluster is exposed to basic pH media, some MO_6 octahedra are removed from the structure, leaving a lacunary-type structure. Lacunary-type structures can be understood as stable intermediates along the pathway of the base-induced decomposition of a POM structure into MO_4^{2-}, during which so-called defects or vacancies remain in the structure – typically one to three vacancies. The remaining defects are reactive and can be filled with a (foreign) element. The intact POM structure type can be regenerated in an acidic pH medium, yielding a (foreign) element substituted POM. This approach is known as the lacunary approach. POMs are sensitive to pH fluctuations because the structure type can change depending on the pH value. The purification of POMs remains a challenge, as unwanted alkali halides are formed as by-products. Separating these two highly water-soluble components is not straightforward. Modern nanofiltration systems address this issue and are garnering increasing interest in the field.

POMs are made up of different components:

- The cation. POMs interact with their respective cation in a defined way. The solubility of the POM compound can be controlled by the choice of cation (inorganic cations – water solubility, organic cations (TBA) – solubility in organic solvents like acetonitrile).
- Heteroelement. The choice of heteroelement in HPA structures can influence the type of POM structure that is formed.
- Ligands. If the oxo ligand in POMs is replaced by other ligands, the type of POM structure can be modified.

- Hydration water. POMs synthesized in aqueous media are isolated as multihydrates.

In the solid-state, POMs exist as different isomers (structural or positional isomers) with different thermodynamic stability. The observed bond lengths are consistent with established trends in coordination chemistry: a shortened bond length between the terminal oxo ligand an the metal center (double bond character) and an elongated length between the oxygen atom of the heteroelement and the metal (higher *trans* influence of the terminal oxo ligand in the ground state). A POM in aqueous solution is exposed to dissociation equilibria (driven by entropy), whose complexity for TMSPOMs increases with higher substitution degrees.

POMs are used in catalysis (homogeneous/heterogeneous). Applications range from acid/base to RedOx and photocatalytic applications. There are promising applications in which POMs are used as catalysts in green chemistry processes in response to the problems of the 21st century. POMs have attracted particular interest for sustainable catalytic applications due to their stability and solubility in aqueous media. This allows reactions to be done in water, eliminating the need for toxic and volatile organic solvents. In combination with biomass—a natural carbon source—valuable platform chemicals (such as FA) can be produced for the chemical industry. These properties have been specifically exploited in the OxFA and ECODS process. In addition to oxidative applications, reductive processes such as HyFo, hydrogenation and CO_2 reduction have also been reported. Another important class of reactions involves cyclization (*e.g.* copper catalyzed azide–alkyne cycloaddition, CuAAC), which plays a key role in synthetic organic chemistry for the synthesis of heterocycles. Another large field of application for POMs is in biomedicine. All mechanisms of action are based on POMs interacting with biomacromolecules (*e.g.* proteins) in a defined manner (electrostatic, coordinative and hydrogen bridges). POMs can induce apoptosis and are able to inhibit enzymes/proteins, the binding of viruses to specific molecules, and the translation of DNA. It is possible to tailor POMs for these applications by controlling the size and shape of the cluster (by selecting the appropriate structure-type), the surface charge distribution (by choosing the structure-type and foreign element substitution), and the RedOx potential/acidity (also *via* structure-type and substitution). In addition, organic molecules can be grafted onto the surface of POMs to increase their solubility in lipophilic media, and POMs can be incorporated into biomacromolecular matrices to improve biocompatibility and reduce toxicity. For biomedical applications, POMs are investigated in the treatment of viral and bacterial infections, as well as in therapies for tumors, diabetes, and Alzheimer's disease. However, the toxicity of transition metals remains a concern, though it can be mitigated by various strategies. The medical applications of POMs are proving to be highly promising – they may play a significant role in future disease treatments.

The properties and applications of POMs presented in this book show that the chemistry of POMs is rich and not just a niche topic in inorganic chemistry. POMs are proving to be promising candidates for future developments and applications.

Abbreviations

POM	Polyoxometalate
HPA	Heterpolyanion/heteropolyacid
IPA	Isopolyanion/isopolyacid
pH	Negative decadic logarithm of the hydrogen ionconcentration
TBA	Tetrabutylammonium
TMSPOM	Transition metal-substituted polyoxometalate
RedOx	Reduction/oxidation
FA	Formic acid
OxFA	Oxidative formation of formic acid
ECODS	Extraction-coupled oxidative desulfurization
HyFo	Hydroformylation
CuAAC	Copper catalyzed azide–alkyne cycloaddition
DNA	Deoxyribonucleic acid

Acknowledgements

I would like to thank the publisher, the Royal Society of Chemistry, for the opportunity to write this book!

Recommended Reading

Please have a look into the following references for an alternative overview of polyoxometalates:

1. M. T. Pope and A. Müller, *Polyoxometalate Chemistry From Topology via Self-Assembly to Applications*, Kluwer Academic Publishers, New York, Boston, Dordrecht, London, Moscow, 2002.

2. J. J. Borrás-Almenar, E. Coronado, A. Müller and M. Pope, *Polyoxometalate Molecular Science*, Springer Netherlands, Dordrecht, 2003.
3. T. Yamase and M. T. Pope, *Polyoxometalate Chemistry for Nano-Composite Design*, Kluwer Academic Publishers, New York, Boston, Dordrecht, London, Moscow, 2004.
4. E. Coronado and C. J. Gómez-García, *Chem. Rev.*, 1998, **98**, 273–296.
5. D.-L. Long, R. Tsunashima and L. Cronin, *Angew. Chem.*, 2010, **122**, 1780–1803.
6. R. Dehghani, S. Aber and F. Mahdizadeh, *Clean - Soil, Air, Water*, 2018, **46**, 1800413.
7. D. L. Long, E. Burkholder and L. Cronin, *Chem. Soc. Rev.*, 2007, **36**, 105–121.
8. M. T. Pope, M. Sadakane and U. Kortz, *Eur. J. Inorg. Chem.*, 2019, **2019**, 340–342.
9. M. Poper and A. Müller, *Angew. Chem., Int. Ed. Engl.*, 1991, **30**, 34–48.
10. D.-L. Long, R. Tsunashima and L. Cronin, *Angew. Chem., Int. Ed.*, 2010, **49**, 1736–1758.
11. M. Ammam, *J. Mater. Chem., A*, 2013, **1**, 6291–6312.
12. Y. Wei, *Polyoxometalates*, 2022, **1**, 9140014.
13. B. Li and L. Wu, *Polyoxometalates*, 2023, **2**, 9140016.
14. U. Kortz, A. Müller, J. van Slageren, J. Schnack, N. S. Dalal and M. Dressel, *Coord. Chem. Rev.*, 2009, **253**, 2315–2327.
15. A. Müller, E. Beckmann, H. Bögge, M. Schmidtmann and A. Dress, *Angew. Chem., Int. Ed.*, 2002, **41**, 1162–1167.
16. J.-C. Raabe, S. Chitnis and M. J. Poller, *Dalton Trans.*, 2025, **54**, 1772.
17. M. Papajewski, J.-C. Raabe, H. Anwari, D. Voß, J. Albert and M. J. Poller, *Inorganics* 2015, **13**, 176.
18. A. Wesner, J.-C. Raabe, M. J. Poller, S. Meier, A. Riisager and J. Albert, *Chem. – Eur. J.*, 2024, **30**, e202402649.
19. J.-C. Raabe, T. Esser, M. J. Poller and J. Albert, *Catal. Today*, 2024, **441**, 114899.
20. J.-C. Raabe, L. Hombach, M. J. Poller, A. Collauto, M. M. Roessler, A. Vorholt, A. K. Beine and J. Albert, *ChemCatChem*, 2024, **16**, e202400395.
21. S. D. Mürtz, J.-C. Raabe, M. J. Poller, R. Palkovits, J. Albert and N. Kurig, *ChemCatChem*, 2024, **16**, e202301632.
22. J.-C. Raabe, F. Jameel, M. Stein, J. Albert and M. J. Poller, *Dalton Trans.*, 2024, **53**, 454–466.

23. T. Esser, A. Wassenberg, J.-C. Raabe, D. Voß and J. Albert, *ACS Sustainable Chem. Eng.*, 2024, **12**, 543–560.
24. J.-C. Raabe, T. Esser, F. Jameel, M. Stein, J. Albert and M. J. Poller, *Inorg. Chem. Front.*, 2023, **10**, 4854–4868.
25. J.-C. Raabe, M. J. Poller, D. Voß and J. Albert, *ChemSusChem*, 2023, **16**, e202300072.
26. J.-C. Raabe, J. Aceituno Cruz, J. Albert and M. J. Poller, *Inorganics*, 2023, **11**, 138.
27. M. J. Poller, S. Bönisch, B. Bertleff, J.-C. Raabe, A. Görling and J. Albert, *Chem. Eng. Sci.*, 2022, **264**, 118143.
28. J.-C. Raabe, J. Albert and M. J. Poller, *Chem. – Eur. J.*, 2022, **28**, e202201084.

www.ingramcontent.com/pod-product-compliance
Lightning Source LLC
Chambersburg PA
CBHW040754220326
41597CB00029BA/4829

* 9 7 8 1 8 3 7 0 7 2 0 3 3 *